T0313156

LEAN
TRANSFORMATION

Cultural Enablers and Enterprise Alignment

LEAN

TRANSFORMATION

Cultural Enablers and Enterprise Alignment

Suresh Patel

CRC Press
Taylor & Francis Group
Boca Raton   London   New York

CRC Press is an imprint of the
Taylor & Francis Group, an **informa** business

A PRODUCTIVITY PRESS BOOK

CRC Press
Taylor & Francis Group
6000 Broken Sound Parkway NW, Suite 300
Boca Raton, FL 33487-2742

First issued in hardback 2019

ISBN-13: 978-1-4987-4336-5 (hbk)

Library of Congress Cataloging-in-Publication Data

Names: Patel, Suresh (Quality management consultant), author.
Title: Lean transformation : cultural enablers and enterprise alignment /
Suresh Patel.
Description: 1 Edition. | Boca Raton : CRC Press, 2016. | Includes
bibliographical references and index.
Identifiers: LCCN 2015048105 | ISBN 9781498743365
Subjects: LCSH: Organizational effectiveness. | Product management. | Quality
control. | Corporate culture.
Classification: LCC HD58.9 .P3825 2016 | DDC 658.4/013--dc23
LC record available at https://lccn.loc.gov/2015048105

Visit the Taylor & Francis Web site at
http://www.taylorandfrancis.com

and the CRC Press Web site at
http://www.crcpress.com

I want to dedicate this book to Mr. K. K. Nair, executive director, Ahmedabad Management Association, who encouraged me to write it after receiving excellent feedback from the delegates at the first-ever Lean Six Sigma three-day course at the AMA in June, 2011.

I also dedicate this book to my dear wife, Pushpa, who had to bear many disruptions and inconveniences without my help, and without her full cooperation, this book would not have materialized.

Contents

SECTION I   Lean Culture Enablers

SECTION II   Principles of Continuous Improvement Process

SECTION IV   Lean Performance Measures and Performance Assessment

Foreword

1. If you believe in a product, do not give it up halfway through. Be on it. And you will succeed one day and the results will be good.
2. Have patience during difficult times. Do not lose your balance, and try to carry the team with you.
3. If it is a new business, plan for 50% more to standby so that you do not have to close the business or run away.
4. There is a lot of scope in manufacturing. The world's emerging economies can become strong in the long run only through the manufacturing base, not the service base. A service base is only temporary. This will not create long-term employment.

Suresh met me a month ago and requested that I write a foreword for his sister books: *The Global Quality Management System: Improvement through Systems Thinking*; *Lean Transformation: Cultural Enablers and Enterprise Alignment*; *The Tactical Guide to Six Sigma Implementation*; and *Business Excellence: Exceeding Your Customers' Expectations Each Time, All the Time.*

When I met Suresh and came to know about his operational excellence experience of more than two decades with multinational corporations such as Eaton Corporation and Fiat Global, a bell rang inside me, and I made up my mind not only to pen the foreword but also to leverage his Spanish language command to boost the performance of one of my South American Chilean units engaged in manufacturing wear-resistant products and material handling for the mining industry.

I knew Suresh well when I invited him to our Kolkata headquarters to spend one week at the Tega home office and the main plant at Joka. It was evident from the feedback report I received from my plant management team that these sister books will clear the "cob webs" and prepare any organization for the journey of continuous quality improvement.

These sister books are unique and comprehensive "how-to-understand and implement" a global quality management system, Lean system, Six Sigma methodology, and business excellence strategy to achieve world-class business excellence. The author has succinctly summarized the business excellence concept and the body of knowledge of this book by illustrating the business excellence pyramid with the foundation of the management systems at the system level, Lean system at the operational level, Six Sigma methodology at the tactical level, and business excellence at the strategy level.

The first book, *The Global Quality Management System: Improvement through Systems Thinking*, starts by paying homage to leading quality "gurus." Having illustrated systems thinking as opposed to the command and control system, the author then stresses the fact that the command and control system can at worst "influence people to behave in ways which dissatisfy the customer and/or suboptimize performance."

The main emphasis of any quality management system is on the process. The first book stresses the importance of the process—its identification, definition, improvement, and control using a "turtle" diagram and its extension to "SIPOC" (suppliers, inputs, process, outputs, and customers) diagrams. The processes discussed include, among others, main business processes such as the human resource (HR) process, the finance process, and the project management process and importantly the "process of improving the process."

Every documented GQMS has a focus on customer requirements and management system processes, which lead to customer satisfaction. To this end, the author has included advanced processes to comply with ISO 9001, ISO/TS 16949, and AS 9100 standards and elaborated on management improvement through extensive plan–do–check–act (PDCA) analysis and the problem-solving methodology involving the famous eight disciplines process (8D). The "check" and "act" phases are extensively discussed through audit processes and a process control plan audit (PCPA) as practiced by most automotive and multinational corporations.

The second book, *Lean Transformation: Cultural Enablers and Enterprise Alignment*, is about the Lean system. Section I explains why Lean implementation usually fails. It goes on to show the approach for Lean transformation by highlighting the "cultural enablers" for the employees (including management) and how management should align the Lean transformation process. In Section II, the book explains principles of continuous process. Section III is about the Lean tools and how they can be deployed for continuous improvement. Section IV is about Lean performance measures and how to assess Lean performance. Assessment of the Lean system tools is a very interesting feature of this part and enables an organization to remain focused on the standardization of the Lean system and to boost the organization's sustainability efforts.

The author has succinctly portrayed the main principles of the Lean system as follows:

1. Define customer requirements correctly and arrive at customer value so you are providing what the customer actually wants.
2. Identify the value stream for each product/service family and remove the non-value-added (wasted) steps for which the customer will not pay and that do not create value.
3. Make the value stream flow continuously to shorten throughput and delivery time aggressively.
4. Allow the customer to pull product/service from your value streams as needed (rather than pushing products toward the customer on the basis of forecasts).
5. Never relent until you reach perfection, which is the delivery of pure value instantaneously with zero waste and zero defects.

The third book, *The Tactical Guide to Six Sigma Implementation*, is about the unique way in which the so-called difficult concept and practice of Six Sigma methodology are depicted. It includes the collection of tools needed for all five phases, define, measure, analyze, improve, and control (DMAIC), and proven best practices to identify which few process and input variables influence the process output measures. To begin with, the author describes the basic concepts of variation, spread of data, and sigma through basic statistical concepts. Before embarking on the five phases (DMAIC), the author clarifies what is needed for business performance measurement through the concepts of "balanced score card" and important measuring units for the quality performance. Notable measures discussed are defects per million opportunities (DPMO), rolled throughput yield, cost of poor quality (COPQ), business failure costs, cost benefit analysis, return on assets (ROA), and lastly, a method of evaluating projects and investments known as net present value (NPV) or discounted cash flow (DCF).

The five phases—define, measure, analyze, improve, and control (DMAIC)—form the bulk of the third book. The step-by-step approach taken by the author to explain the key concepts and tools required in each one of these phases requires a special mention.

Define phase: This phase starts with the definition of the voice of the customer (VOC). Here the quality function deployment (QFD) tool is described in a simple and easy way to translate the customer's voice into the language of the engineer. The QFD is then used to define and document a business improvement project charter based on the customer and competitive intelligence data. The project tracking tools such as Gantt chart, critical path analysis (CPA), and project evaluation and review technique (PERT) are explained in detail. The CTQ flow down is introduced to define the customer satisfaction in four areas: quality, delivery, cost, and safety for internal and external customers.

Measure phase: The author has identified and discussed 16 different aspects of process characteristics. Having done this, a measurement system analysis (MSA) is described in great detail to ensure that the integrity of the measured data of important characteristics, the measuring equipment, and the human aspect of the measurement system is maintained within allowed repeatability and reproducibility (R&R) acceptance criteria.

Analyze phase: Here the root cause analysis methods for the problems encountered are discussed. The main techniques described include regression and correlation, ANOVA, FMEA, gap analysis, waste analysis, and Kaizen.

Improve phase: The process improvement methods discussed in this phase are prioritization through the cause and effect (C&F) matrix, Kaizen using Lean tools and Six Sigma, PDCA, and theory of constraints.

Control phase: The key concepts and tools illustrated in the control phase are statistical process control (SPC), TPM, and total productive maintenance and overall equipment effectiveness (OEE), MSA, control plan, and visual factory. To sustain the improvements, the tools referred to are as follows: lessons learned, training plan, standard operating procedures (SOPs), work instruction, and ongoing performance assessment.

Design for Six Sigma (DFSS): This methodology is a very useful and logical extension of the Six Sigma phases. The tools discussed in DFSS include define, measure, analyze, design, and verify (DMADV) and define, measure, analyze, design, optimize, and verify (DMADOV). Design for X (DFX) includes reliability analysis and design of tolerance limits. Special design tools described are Porter's five forces analysis and TRIZ (Russian for "theory of inventive problem solving").

Any Six Sigma book cannot be called complete without a case study. To this end, the author has chosen an improvement project to improve the batting in the cricket game using the Lean Six Sigma approach.

The fourth book, *Business Excellence: Exceeding Your Customers' Expectations Each Time, All the Time*, is about business excellence strategy. There are many models of business excellence practiced by many countries. At best, these models lay down business excellence assessment criteria, but the author has felt that the main requirement of the organizations intending to embark on business strategy needs a special body of knowledge with which the business excellence strategy can be implemented successfully. The inclusion of strategies for leadership, strategic planning, customer excellence, operational excellence, and functional excellence for HR and IT will prove to be very useful for the initiated management. Assessment of business excellence strategy through the use of the balanced score card, employee survey, achieving performance excellence, and cost out is a very effective chapter for the business excellence strategy.

Finally, as you will put these sister books of knowledge into practice, you will discover the shifting roles of leaders and managers in your organization. It is not enough for the leaders to just keep on doing what they have always done. It is not enough for them to merely support the work of others. Rather, leaders must lead the cultural transformation and change the mind-sets of their associates by building on the principles behind all these excellent tools.

The author's account of these difficult and vast subjects is very praiseworthy and proof of his vast industrial experience over four decades working with MNCs in Asia, Europe, and the Americas. This is an inspirational work easy to be learned and applied by the lay reader. I highly recommend this book to all those students, teachers, and executives and organizations who want to learn and implement GQMS Lean Six Sigma systems and business excellence strategies.

Madan Mohanka
CMD and Founder
TEGA Industries Limited
Kolkata, West Bengal, India

Foreword

The *Quality and Business Excellence* sister books create and deploy the preventive quality culture within an organization. These sister books enhance customer value and satisfaction by fully integrating the customer's voice into the design, manufacturing, supply chain, and field processes.

Almost all business organizations are engaged in providing services or products to their customers. However, when it comes to providing service to customers and presenting them an experience that will make them come back time and time again, only a small minority of organizations stand out from the crowd who apply the energy, commitment, and innovative thinking to get it right. There is an enormous difference between those who are truly focused on customers and those who simply pay lip service.

These sister books have been written primarily for business entrepreneurs, business managers, engineering managers, and technocrats who wish to grasp Global Quality, Lean, Six Sigma, and Business Excellence concepts, methodologies, and tools to improve and to promote their companies to world-class standards for profitability and sustainability.

These sister books can be used as a basic textbook for a Green Belt, Black Belt, BBA, and MBA courses in Global Quality, Lean, Six Sigma, and Business Excellence.

The word "quality" has been applied in many enterprises, mostly by quality professionals and consultants. Lately, the word "quality" is replaced by "continuous improvement." These two words have become "continual improvement" in ISO 9000 standards, and now have finally become "continuous quality improvement." The subsequent proliferation of terms tends to confuse managers in the marketplace. ISO9001, ISO/TS16949, AS 9100, JIT, MBNQA, Six Sigma, Kaizen, Kanban, 5S, Lean, TPM, TQM, etc., are only a few of the initiatives confronting organizational leaders. No wonder managers are confused. Too many consultants are trying to sell the next "fad" or "savoir" to gain an advantage in the marketplace.

These sister books clear the cobwebs. It prepares the initiated person/organization for the journey of continuous improvement. The guiding principle is

> An organization must constantly measure the effectiveness of its processes and strive to meet more difficult objectives to satisfy customers.

Taiichi Ohno
Toyota Production System

Acknowledgments

Acknowledging the help and guidance in writing these sister books is like churning the oceans of the world and putting all the blessings in a teacup. I find it very daunting because during my more than 50 years of industry experience, I have been guided and helped by many persons, companies, and institutions with whose associations, I have learned, practiced and taught these subjects and achieved modest to excellent results.

After I decided to return to Ahmedabad from Texas, Prof. R. D. Patel, a finance professor at IIM Ahmedabad, asked me to address their SME program as a guest speaker to talk about Lean Six Sigma. The feedback from the attendees was good, and he (Prof. Patel) took me to the Ahmedabad Management Association (AMA) to meet with its executive director, K. K. Nair, who asked me to conduct a second five-day AMA Lean Six Sigma seminar attended by industry representatives from Rajkot, Vadodara, Surat, and Ahmedabad. This led to another seminar at AMA and an invitation by the HR head of ISRO (equivalent to Indian NASA), J. Ravisankar, to address ISRO technicians and engineers on the subject of zero defects delivery of space systems, which was well received.

Consequently, K. K. Nair asked me to write a book on Lean Six Sigma for Indian engineers. My learning and experience as an operations excellence and engineering manager at Eaton Corporation (Eden Prairie, MN, USA) and Fiat Global (Burr Ridge Operations, Chicago, IL, USA) made me take a holistic view and include the global quality management system at the bottom rung and business excellence at the top level. So I wound up writing four books.

I thank the following individuals for their contributions to my knowledge and all the help and guidance they offered me in my career, which resulted in creating these books: C. S. Patel, former CEO of Anand Group, leading automobile companies manufacturing automotive components; the late D. N. Sarkar, CMD of Gestetner Limited; Samir Kagalwala, consultant for the design and manufacture of power magnetics; Stefan Lorincz, renowned electronics engineer and source developer for key electronic components worldwide at Phillips, Holland; Levy Katzir, former Motorola VP, who put me in charge of the quality and reliability of the newly developed electronic ballasts in 1994; G. P. Reddy, former director of quality at Universal Lighting Technologies; Inder Khatter, international QMS lead auditor for DNV, Houston, USA; Dev Raheja, international consultant and co-author of *Assurance Technologies Principles and Practices*; Frank Kobyluch, global general manager at Klein Tools and former plant manager at Eaton Corporation; and Don Johnson, director of quality at Fiat Global–Case New Holland Division.

My special thanks and gratitude also go to my colleagues and team members at the following companies where I worked, learned, developed, and implemented many of the tools and techniques contained in these books: Gestetner Limited (now Ricoh India); Energy Savings, Inc., Schaumburg, Chicago, IL; United Lighting Technologies, Nashville, TN, USA; Eaton Hydraulics, Eden Prairie, MN; and Fiat Global–Case New Holland, Burr Ridge, Chicago, IL.

My abilities as an operations excellence manager in charge of providing quality products for CFL ballasts, hydraulic valves, pumps, hydraulic hoses, and fittings were honed, tested, and appreciated by customers such as GE CFL Lamps, Osram Sylvania, John Deere, Case New Holland, Oshkosh Corporation, manufacturers of severe heavy-duty all-wheel drive defense trucks, Caterpillar, GM trucks, Ford trucks, Volvo trucks, Zamboni ice resurfacer (for Olympic Games), etc.

I will remain grateful to the following suppliers who collaborated with me and my team in developing components and major assembly units requiring extremely high-precision and pre- or posttreatments: Parker Hannifin, which supplied high-quality hydraulic seals and O-rings; Bosch, which supplied specialty hydraulic valves; Carraro Pune, which supplied a complete four-speed transmission unit for agricultural tractors; TGL-Carraro Pune, which developed precision gears and shafts for transmissions; Carraro, Quingdao, China, with which we developed the entire rear axle assembly for backhoe loaders; Graziano Transmission, Noida, India, in which we developed a continuously variable transmission unit for a tractor for the first time for the U.S. market; GNA Group Punjab, which supplied us forged and precision machined components for the tractor transmission assemblies; and Craftsman Automation Limited, Coimbatore, which machined our large castings for transmission bodies and covers using heavy CNC machines and digital CMMs.

I have remained in touch with developing technology and professional knowledge through the American Society for Quality, whose membership I have had since 1993.

Illustrations and design of charts and figures in these books were done by Sanjay Trivedi and Minal Mehta.

Making It Big in Manufacturing Product and Providing Service

It is a general belief that successful people in every field are blessed with talent or are just lucky. However, the fact is successful people work hard, work long, and work smart.

Marissa Ann Mayer, the current president and CEO of Yahoo, used to work 130 hours per week while working at Google. India-born Indra Krishnamurthy Nooyi, the chairman and chief executive officer of PepsiCo, worked midnight to 5 a.m. as a receptionist to earn money so that she could complete her master's degree at Yale University. In 1958, Qimat Rai Gupta left his education midway and founded Electric Trading operations in the electric wholesale market of Old Delhi, India. With an investment of Rs 10,000 ($150), he started Havells Industries. Today, Havells is a billion-dollar company. In his own words, "Overnight success means 25 years of hard work, devotion, and dedication."

The story of the founder and CEO of the Kolkata, India–based Tega Industries, Madan Mohanka, is unique. When he went into business, he had the right combination—hailing from a business family, having an engineering degree, getting an MBA from IIM Ahmedabad, and having a foreign collaborator as a joint partner. Yet this combination failed miserably. He was witnessing the imminent closure of his company in 1979, but like the epic hero Odysseus, he never lost focus for a moment. He kept at it. Some three decades later, it is Madan's die-hard optimism that saw Tega Industries become the second-largest player in the world in rubber mill lining products for the mining industry.

In her book *Stay Hungry, Stay Foolish*, Rashmi Bansal (IIMA Graduate) depicted Madan Mohanka's hard-won story very aptly. She said Madan faced all hurdles and challenges starting from scratch, but then Madan had what you call an "obsession." Over the last three decades, Madan built a strong foundation combining three technologies, viz., mechanical engineering, rubber (polymer) technology, and mineral processing and grinding. Over the last 5 to 6 years, Tega accepted challenges, grabbed overseas marketing opportunities, and maintained consistent growth keeping an eye on the margins.

Tega's presence in 19 international locations has enabled it to increase a turnover of $4 million in 2009 to $120 million in 2014.

According to Mehul Mohanka, Madan's son who trained and earned an MBA in the United States, the stage is now set for organic and inorganic growth—organically building up larger capabilities and inorganically looking for acquisitions for successful integration with Tega's culture, values, and philosophy.

Section I

Lean Culture Enablers

1

Introduction

Many people often come across the word *Lean* and wonder what it is all about. Well, Lean is neither about getting a six-pack belly nor about being skinny.

Lean is a methodology for operational excellence. It is based on the Toyota Production System (TPS). TPS differentiates between two fundamentally different business systems—Lean production versus mass production. These are two very different ways of thinking to create value for the customer.

"Lean manufacturing" is a favorite buzzword of business circles today. It is easy for the uninitiated to be unclear on the difference between Lean and mass manufacturing. In fact, they differ in many ways, from philosophy and business strategy to production models and company culture (Table 1.1).

Lean is defined as follows:

- A passionate belief that there is always a simpler, better way; work smarter not harder
- A continuous drive to identify and eliminate waste
- A way of thinking that empowers employees (including executive employees) to use their talent to improve the business every day
- A culture that extends the concepts of Lean across all business processes using a common tool set

TABLE 1.1

Main Characteristics of Mass and Lean Production

Lean Production	Mass Production
Low volume	High volume
Many products	Few products
Rapid setup	Slow setup
Short product life cycle	Long product cycle in development, design, and manufacturing phases
Fast response to market change	Slow response to market change
Control based on product knowledge	Control based on part numbers

HISTORY OF LEAN

For the following compilation of the history of Lean, the main reference is from Womack, Jones, and Roos (1990, *The Machine that Changed the World: The Story of Lean Production*, New York, HarperCollins).

Early 1900s

Henry Ford creates efficient assembly lines—mass production. The Model-T automobile is considered as the first example of Lean production with simple design and interchangeable parts.

1913

Ford implements a continuous-flow (moving) assembly line, slashing cycle times. Ford separates skilled trades from assemblers to speed up training.

1920

Ford produces more than 2 million vehicles per year and cuts costs by two-thirds.

1920s

Ford Motor Company's operation adopts the key element of scientific management promoted by Frederick W. Taylor (1856–1915) having the following main characteristics:
- Standardized product designs
- Mass production
- Low manufacturing costs
- Mechanized assembly lines
- Specialization of labor
- Interchangeable parts

1926

Toyoda Loom Works develops Jidoka—a device that detects a broken thread and stops the loom. One operator can thus operate many looms.

1937

Toyota Motor Company (TMC) is born (looms and trucks for military and finally automobiles).

1940s

Taiichi Ohno, Toyota's chief production engineer, experiments with U.S. presses and perfects a quick die changeover called "single minute die exchange" (SMED).

1950

Eiji Toyoda, a Japanese engineer, visits Ford Motor Company with Taiichi Ohno, beginning the Lean manufacturing revolution. The 13-year effort of Toyota Motor Company had produced 2685 automobiles up to 1950, compared with the 7000 automobiles produced daily at the Ford Rouge Plant in Detroit. This was soon to change (Womack et al., 1990, p. 48).

Early 1950s

The U.S. automotive is mighty—the United States and Europe embrace mass production and automation (Ohno sees waste).

1950s–1960s

The Lean manufacturing revolution is born. While the United States focuses on mass production, at TMC, Ohno institutes defect prevention, teamwork, problem signaling, pull production, flow control, small lot sizes, and supplier integration.

1961

Unimate is the first robot in production in the United States—part of the automation revolution, human replacement concept.

Early 1970s

Microprocessor technology is developed—programmable automation is made possible.

1973

Fuel prices increase dramatically. Consumers' preferences change: They want fuel efficiency and compact size. Ohno and Eiji Toyoda's 20-year focus on productivity, quality, and responsiveness (just in time [JIT]) comes into play.

1980s

With quality revolution and Six Sigma born, Toyota grows in popularity, and U.S. automobile manufacturers begin to see a negative shift. The United States focuses on technology (automates waste at record rate).

1988

International Motor Vehicle Program (IMVP)/MIT researcher John Krafcik coins the term *Lean* in his research article "Triumph of the Lean Production." He concludes, "Lean production uses half the human effort, space, tools, engineering hours to develop new products as compared to mass production. Lean production has less inventory, fewer defects, and produces greater variety of products." Later on, he joined Ford Motor Company, and now he is with Hyundai Motors.

Late 1980s

Toyota is a legitimate threat; fuel efficiency and customer preferences are critical success factors. JIT and total quality management (TQM) become important to the United States.

1990s

Customer focus (customer is always right) is observed in the global marketplace: Outsourcing and international trade are fully embraced, and TQM becomes more emphasized.

1990s

With the evolution of Six Sigma/DMAIC (Motorola), Lean manufacturing significantly grows in popularity, but implementation is difficult.

2000

With Lean Six Sigma, Lean is applied to fields other than manufacturing (e.g., construction).

Present

TMC is the world's largest automobile manufacturer.

Lean is applied to all fields, service industries, supply chains, and business processes, e.g., Lean office, Lean healthcare, Lean farming, Lean construction, Lean accounting, and Lean graphic communications.

Today, employers want people who are motivated, trainable, educated, and skilled (but with operations management abilities).

Today, efficiency is the primary driver, not technology.

2

Business Process

ROADBLOCKS TO LEAN TRANSFORMATION

Most organizations want quick fixes and immediate results. If they do not get the immediate reward they seek, they may abandon the program, the team, or go in search of the next prevailing fad. This need for short-term results has caused the total discard or partial success of many performance excellence programs like Lean transformation.

Successful transformation to a Lean culture is full of hardships. It requires an enterprise-wide approach that engages the entire organization and challenges its norms and existing practices. It requires knowledge of new tools and methodologies, and a level of internal winning instinct beyond the existing discipline and willpower in which the majority of organizations operate.

Why does Lean work in some organizations and not in others? In short, the difference between success and failure is in cultural acceptance and the ability of an organization to accept change, not just Lean change, but change in general. Understanding the mind-set of business is crucial to the success of implementing methodologies like Lean and Six Sigma.

Let us first understand the business system. As we discussed earlier, a system is "a group of interacting, interrelated, or interdependent elements forming a complex whole." No system element can function on its own. It has to rely on other elements and has to maintain its relationship with other elements.

The human body is a system where hands, feet, stomach, heart, etc., are the elements enabling the body to function as a whole. No element can function on its own. Dr. Ackoff once joked, "Try cutting off your hand and put it on the table—it won't work!"

Now, let us apply the system concept to business and as stated in the beginning; how does this translate into implementing Lean and Six Sigma for a business?

Let us consider a business having the following core functions:

- Sales
- Marketing
- Engineering
- Production
- Customer service

These core business functions are depicted in Figure 2.1.

Each of these core functions has its own set of defined processes, as shown in Figure 2.2, and it uses its processes to accomplish its goals. Now let us recap process definition. Process is defined as "a series of actions, changes, steps, or functions bringing about a result."

In Part 1, we explained that each process has the following elements that affect its function:

- Inputs
- Process
- Outputs

Now, apart from the main core functions (Figure 2.2), business is also supported by typical support functions like:

- Human resources
- Finance
- Information technology (IT)
- Warehousing

These support functions are depicted in Figure 2.3.

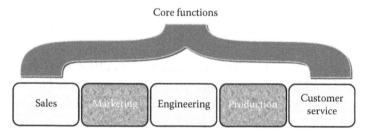

FIGURE 2.1
Core business functions.

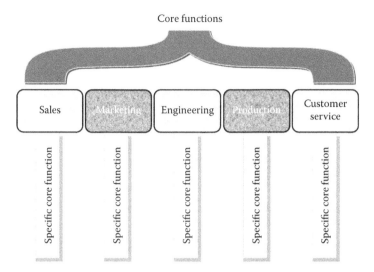

FIGURE 2.2
Each core function has its own processes.

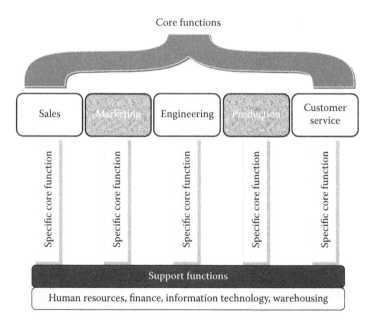

FIGURE 2.3
Businesses and support functions.

At this point, the analysis of the business system looks vertical (Figure 2.3). Individuals inside a particular functional area have full view of their own process but have difficulty seeing outside of these "silos." They intersect with another functional area only when they need to use a common resource. For example, an intersection between sales and production functional areas occurs when a tracking system managed by the information technology support function is used by the production function to deliver a product to a customer. In short, each individual part of the business system is trying to work on its own. Another example of this silo mentality can be commonly seen with engineering when engineering changes are carried out without changing production processes.

This limited perspective is why it is crucial to understand the business processes that cut across these functional process areas.

BUSINESS PROCESS VIEW

A business process is a collection of related activities that produce a product or service of value to the organization, its stakeholders, or its customers.

Let us look at the following examples of business processes:

- Quote-to-cash
- Procure-to-pay
- New product/service development
- Order fulfillment
- Process impact on the organization

Becoming familiar with these cross-functional business processes described in Figure 2.4 greatly increases our understanding of the interrelationships between the core functions and clarifies how a quality project in one area of the company will affect other areas (Figure 2.4). This interaction and interdependence among core functions is the key to removing the

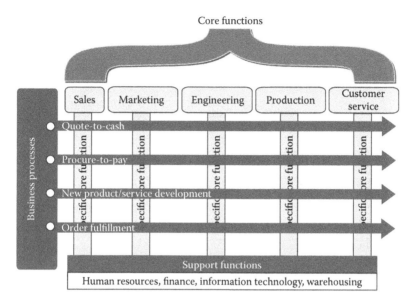

FIGURE 2.4
Cross-functional business processes.

roadblocks in implementing Lean Six Sigma. To truly grasp the system, how-ever, we must consider another aspect of the business process: Its purpose.

MANAGING THE PURPOSE

No business process can be effective unless its purpose is properly com-municated to all stakeholders. Figure 2.5 shows the main purpose of the organization. Executive leadership should drive management of the business purpose, and impress upon all members of the organization the importance of understanding and fulfilling that purpose. In addition, leadership must govern, manage, adjust, and reset the purpose based on the customer's needs and other factors.

In the *Global Quality Management* book, we have seen that there are many input, output, and feedback processes for an organization. All inputs and outputs of a particular process should be measurable so that quality can be controlled. See Figure 2.6.

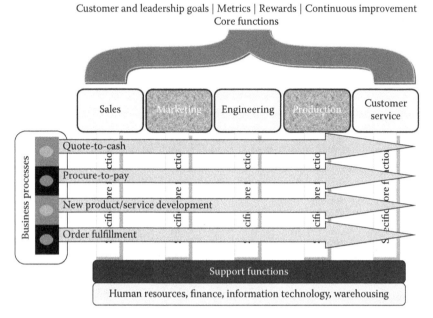

FIGURE 2.5
Purpose of the business is to manage goals and improve continuously.

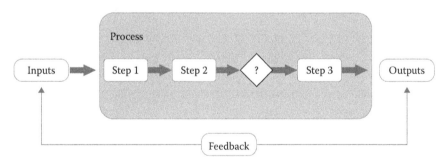

Conduct measurements at both points to gauge
the efficiency and effectiveness of the process.

FIGURE 2.6
Process inputs and outputs with feedback and measurement points.

SIPOC DIAGRAM

Suppliers, inputs, process, outputs, and customers (SIPOC) diagrams are a tool that can be used to help identify these processes in an organization. We have discussed SIPOC in more detail in Part 1, but it is important to know that improvements in one area may create errors in another. For example, if the business purpose of improving profitability is not made clear to sales function, it may inflate its sales forecast at the expense of too much overtime expense, excess work in process inventory, and high logistics price for production function.

In this chart (Figure 2.7), "how," "with what," "with whom," and "how often–how much" goals, etc., can be added as shown in the process identification chart (turtle chart) in (Figure 2.8).

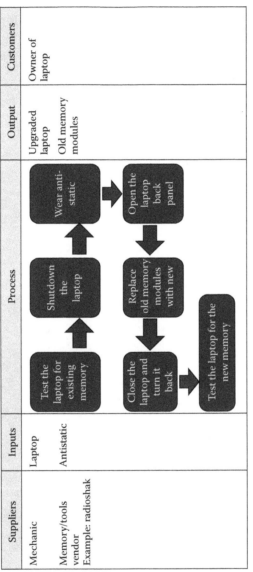

SIPOC diagram for the example process of "upgrading laptop memory."

FIGURE 2.7
SIPOC diagram.

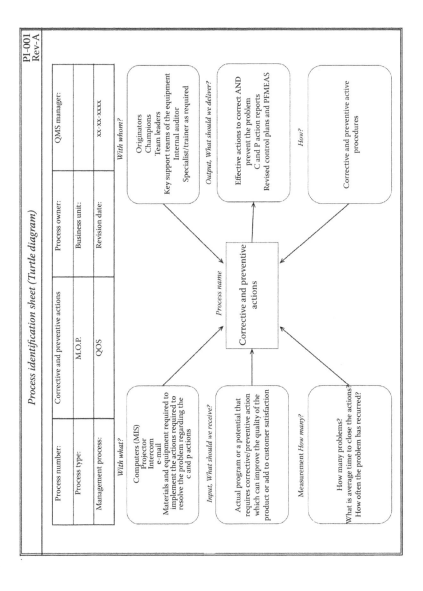

FIGURE 2.8
Turtle chart.

3

Embarking on the Lean Journey

SHINGO PROCESS

Let us start the Lean journey. In the book *Good to Great* (Collins Business, 2001), the author Jim Collins talks about the flywheel effect. Great organizations do not become that way overnight. He writes, "They become that way because continuous small pushes create a breakthrough velocity. This velocity sustains growth. At the point where the momentum of change reaches breakthrough velocity (the tipping point), the organization moves forward along its Lean journey."

An organization normally contains a majority of the "anchor draggers," few "early adaptors," and "fence sitters." While traditional managers spend their efforts focusing on the anchor draggers, they should be spending time with the early adapters, providing cover and support. The organization's focus should be on positive reinforcement, which promotes a forward shift in the fence sitters. When we talk about the best roadmap to implement "Lean," we should use the slide presented in Figure 3.1, which describes the guiding principles and how to support them.

The people aspect, called *cultural enablers* by Shingo, is often missed when organizations say they are doing Lean.

Mark Graban, a leading Lean author and practitioner, has coined an acronym—LAME (Lean as executed mistakenly or Lean as misguidedly explained)—for such attempts.

Now refer to Table 3.1. Let us talk about the difference in style of management that will be required by leadership in a Lean organization.

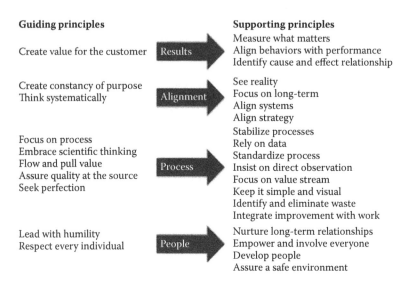

FIGURE 3.1

The guiding principles and how to support them.

TABLE 3.1

Control and Command versus Lean Management

Control and Command	Lean Management
All knowing	Humility
"In charge"	Curiosity
Autocratic	Facilitator
"Buck stops here"	Teacher
Impatient	Student
Blaming	Communicator
Controlling	Perseverance

WHITE COAT LEADERSHIP VERSUS IMPROVEMENT LEADERSHIP

On the left-hand side, we see behaviors and beliefs that are pretty common in most organizations and most leaders. They come from the way people were taught (in school or on the job), and people who were managed in this way were rewarded and recognized for good management. It is the way we saw our bosses manage. "The system" produced these behaviors. (A system is a "set of interdependent components working together toward a common aim"—that definition, from W. E. Deming will come up again later.) (See Table 3.1.)

The list on the right-hand side is not just a little different—it is a lot different. Leaders need to exhibit humility (admitting they do not know it all); they need to be curious about how things work (because they do not know), they need to become facilitators not "tellers," and they need to become teachers. However, before they can teach, they need to know something—they need to be students.

This list is similar to a list that you will find in Dr. Deming's 1993 book *The New Economics* (see Table 3.2). On the left-hand side, we see what Dr. Deming called "the prevailing style of management"; on the right-hand side, "better management practice" presents practices of management with suggestions for better practice (W. E. Deming, *The New Economics*).

The point to note is that moving from the left-hand side to the right-hand side requires the cultural transformation of management. The following is a quote from *The New Economics* by Dr. Deming:

> The first step is transformation of the individual. This transformation is discontinuous. It comes from understanding the system of profound knowledge. The individual, transformed, will perceive new meaning to his life, to events, to numbers, to interactions between people. Once the individual understands the system of profound knowledge, he will apply its

TABLE 3.2

Cultural Transformation from Present to Better

Present Management Practice	Better Management Practice
Reactive skills only required	Lean management theory required
Short-term thinking; failure to optimize through time	Do long-term planning through constancy of purpose
Reward at the top, punishment at the bottom. Through a so-called merit system, rank people, teams, and divisions	Abolish merit system; run the company as a system
Incentive pay	Abolish incentive pay and pay based on performance
Failure to manage corporation as a system; components are treated as individual profit centers	Manage corporation as a system; encourage communication and continual learning
Management by objectives (MBO)	Management by facts (MBF)
Setting numerical goals	Work on methods of improvement of processes

principles in every kind of relationship with other people. He will have a basis for judgment of his own decisions and for transformation of the organizations he belongs to.

The transformed individual will

- be a role model;
- be a good listener, but will not compromise on Lean principles;
- continually coach other people;
- help people to detach from their current practice and beliefs and move into the new philosophy without a feeling of guilt about the past.

If an organization on the left-hand side realizes that the prevailing style of management is not going to be sustainable, they may wish to move to the right-hand side.

If an organization on the left-hand side realizes that the prevailing style of management is not going to be sustainable (costs are higher than revenues, not producing results that provide value to customers), they may wish to move to the right-hand side.

But how? They build a bridge to close the gap. This is called the *infrastructure*, and there are lots of ways to do this—lots of types of "bridges."

Hence, they start to mobilize across this first bridge by trying to learn about Lean thinking and methods. They often do so by training people how to use various Lean tools, using teams in organized events, and they do produce results—at least for a while.

They have trained people and learned Lean tools. However, the rate of improvement decreases over time, and they are still not across this deep gap. Their style of management has not transformed. Hence, what do many organizations do?

Some couple their Lean efforts with Six Sigma, and this reenergizes the effort—for a while. However, it too falls short, and the improvement is not sustained.

Some organizations try this again with "Lean and theory of constraints (TOC), Lean and green, etc.," but it too falls short.

Thus, they are disappointed and perhaps find themselves hanging on or wanting to scurry back to the left-hand side (current state). They claim this "Lean stuff" does not work, and management starts looking for the next "shiny object."

It is a common experience for many organizations that start with events, tools, and training, thinking that this will be sufficient. However, getting trained in Lean tools will not get them all the way across the deep gap.

Figure 3.2 shows that focusing on tools will get temporary results. See the right-hand image called the *Shingo transformational process*. The half portion below the bold line shows that the application of tools will get the results. It is not that the approach is wrong; it is just incomplete. It is a necessary part of learning. Thus, what else is needed to complete the transformation? This is where the "diamond" (or transformation element) of the Shingo model can help.

Through personal and organizational understanding of the 10 guiding principles, the design and the redesign of systems by adjusting the tools that are used, the organization not only achieves sustained results but also impacts behaviors to achieve a true cultural transformation. This is described in more detail in Figure 3.2.

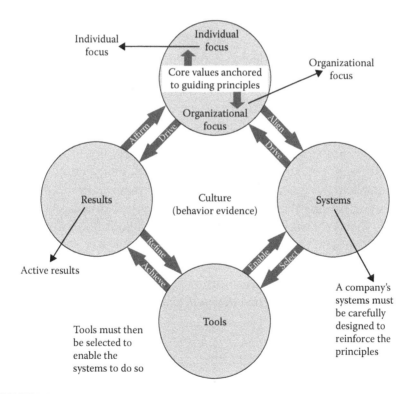

FIGURE 3.2
The "diamond element"—individual and organizational focus. (Courtesy of Shingo Prize.)

Where did these 10 guiding principles come from? Figure 3.3 shows that over the years, there have been many thought leaders who have contributed to these guiding principles.

These principles govern consequences for everyone. The principle of gravity governs us regardless of what we believe. In the same way, the guiding principles such as "respect for every individual," "focus on process," or "constancy of purpose" govern us.

NURTURING THE LEAN CULTURE

Success or failure of a new initiative such as Lean depends entirely on people. There can be no Lean culture without people. Lean tools, Lean equipment, and Lean consultants are of secondary importance.

Thus, we can embrace the Lean culture through the following:

- Respect for the individual
- Leading with humility
- Techniques and practices to change to the culture

Respect for the Individual

"Respect for the individual" is a little harder to define. Lean is not about "being nice" and smiling all the time. Respect means you select and recruit the right people. Train them to become skilled, and then hold people accountable to the system, following it and improving it. Respecting individuals and treating them with dignity is the most critical aspect of Lean culture. Respect starts with top management and flows down through the organization right up to the individual worker or operator.

However, respect for the individual does not end within the four walls of the organization; it extends to all customers, suppliers, and stakeholders.

Through respect, the organization will achieve the following goals:

- Promote team work
- Improve employee involvement
- Empower employees
- Create the culture of continual improvement
- Redeployment of persons

"*Principle*" thought leaders

W. Edwards Deming	Kiichiro Toyoda	Stephen R. Covey	Jim Womack
1. Create constancy of purpose for improving products and services. 2. Adopt the new philosophy. 3. Cease dependence on inspection to achieve quality. 4. End the practice of awarding business on price alone; instead, minimize total cost by working with a single supplier. 5. Improve constantly and forever every process for planning, production, and service. 6. Institute training on the job. 7. Adopt and institute leadership. 8. Drive out fear. 9. Break down barriers between staff areas. 10. Eliminate slogans, exhortations and targets for the workforce. 11. Eliminate numerical quotas for the workforce and numerical goals for management. 12. Remove barriers that rob people of pride of workmanship, and eliminate the annual rating or merit system. 13. Institute a vigorous program of education and self-improvement for everyone. 14. Put everybody in the company to work accomplishing the transformation.	1. Challenge Always be faithful to your duties, thereby contributing to the company and to the overall good. 2. Seek perfection Always be studious and creative, striving to stay ahead of the times. 3. Direct observation Always be practical and avoid frivolousness. 4. Respect Always strive to build a homelike atmosphere at work that is warm and friendly. 5. Humility Always have respect for God, and remember to be grateful at all times.	Independence to interdependence • Habit 1: be proactive—principles of personal choice • Habit 2: begin with the end in mind—principles of personal vision • Habit 3: put first things first—principles of integrity and execution • Habit 4: think win–win—principles of mutual benefits • Habit 5: seek first to understand, then to be understood—principles of mutual understanding • Habit 6: synergize, principles of creative co-operation Continuous inprovement • Habit 7: sharpen the saw—principles of balanced self-renewal of body	Specify value Identify value Stream Flow Pull Seek perfection

FIGURE 3.3

Deming's 14 points, Toyoda's continuous improvement process, and Stephen Covey's seven habits, Jim Womack's Lean approach, and 10 guiding principles are like laws of physics.

A Big No–No

The Lean program will certainly fail if employees who are no longer needed in the new improved process are let go. A common philosophy of Lean-minded employers should be that "no one will lose his or her job as a result of our continual improvement program." So what do you do with the extra people made available after the process improvement?

You can

- reduce temporary workers if applicable,
- reduce overtime,
- absorb redeployed employees to fill in vacancies created because of normal attrition,
- provide for growth (growth is assumed because of Lean implementation) without hiring additional persons,
- retrain and redeploy to where demand is required.

WIFM Resolution (What Is in It for Me?)

In addition, employees ponder over the question What's in it for me after deploying Lean and eliminating waste? The response is an effective communication explaining the following:

- A secure future in a financially sound, competitive company
- An opportunity for excellence
- A voice in how things are done (a feeling of involvement)

Leading with Humility

Lean leadership involves, enables, and empowers people. Lean leadership promotes professional and personal growth of the people. It allows people to take pride in their work. Lean leaders do not set goals for people, sit in their offices, and shout at people when they do not achieve those goals. Lean leaders spend time coaching people. They spend most of their time in the *gemba*. They observe processes with people and see what is actually happening. They do not just manage metrics and read reports.

In Stephen Covey's "The 7 Habits of Highly Effective People," habit 5 says, "To seek first to understand and then to be understood." Instead of jumping in with a desire to be heard, by understanding the other person's point of view, you create better relationships and find valuable solutions to problems. This is personal humility—being humble with self-respect and dignity. Humility enables you to understand your strengths and weaknesses. Strength allows you to become a better team member.

The weakness or the opportunity to improve enables you to continually develop as a successful person.

4

Techniques to Change to Lean Culture

Knowing Lean tools and techniques is *less* than half the job done toward getting transformed to a Lean culture. In this section, we will learn about the ways and practices to instill this new culture through training, coaching, and communication.

- Cross-training
- Skill development
- On-the-job training (OJT)
- Coaching and mentoring
- Teamwork
- Suggestion schemes
- Safe working environment

CROSS-TRAINING: A WIN–WIN SITUATION

- Basically, cross-training consists of training an employee to do different organizational activities or work. Alternatively, cross-training also prepares multiple employees to do a single activity, duty, or work. This leads to a multifunctional workforce.
- Cross-training reduces bottlenecks. Because of their flexibility and ability to carry out multiple functions, employers are ready to face situations like absenteeism, cover for breaks, illnesses, and vacations. The variations in supplies like late and nonconforming deliveries, and customer demands, such as changes in delivery schedules and canceled or changed quantities, are handled efficiently because the organization is capable of adapting to new situations.

- The line supervisors, managers, and even area managers undergo cross-training. This makes them capable of training and filling in for absent employees.
- One big advantage of cross-training is that it promotes continuous improvement. With job rotation, employees are able to observe bottlenecks and problems with fresh eyes and are able to ask questions and suggest remedies.
- Cross-training enables the employees, the company, and the customers to *win*.

STEPS TO IMPLEMENT CROSS-TRAINING

- Communicate the advantages and opportunities it will create for employees and leaders.
- Identify who is interested in getting cross-trained. It is useless to force someone who is unwilling; instead, find out how to deal with such employees.
- Identify existing employee competencies in tasks.
- Develop a training program. Training the trainers is the first step. Either the supervisor or the current capable operator performing that job can impart the training after getting trained as a trainer. For more on "Train the Trainer," visit: http://www.isohelpline.com /train_the_trainers_skill_training_process_powerpoint_ppt.htm.
- Allow the trainees to learn new jobs without undue pressure to complete existing jobs. Allow sufficient time to learn the new job. Two hours per week set for skilled jobs is a good norm.
- Prepare a visual cross-training record and display it for everyone to see. Here, the employee (preferably with photo) and job matrix is created such that the current status of each employee is seen by all. Each employee status ("not trained," "in training," or "trained") for every job can be seen on this wall chart (Figure 4.1). This promotes pride among employees.
- Lastly, recognize and reward the employees who gain new skills and responsibilities.

For every task in an organization, determine the skill set required. This information can be gathered from work instructions, procedures,

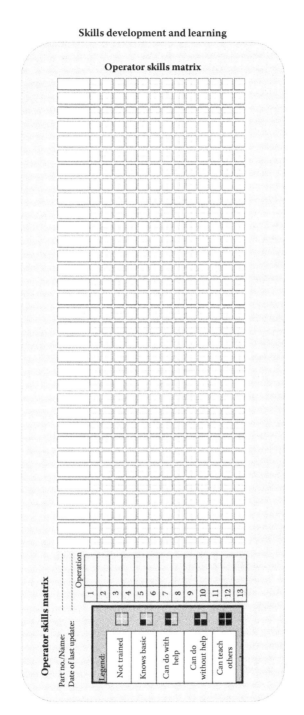

FIGURE 4.1
Operator skill matrix.

test methods, quality plans, process experts, and experienced operators (Figure 4.2). Skills such as welding, soldering, riveting, testing, assembling using torque applicators, data entry, among others.

List all employees in a department who are appointed to do that complete task or assembly. List the skills required needed by that department. Refer to the skills matrix and using the guide shown in Figure 4.3, complete the matrix.

This skill matrix needs to be updated and reviewed each time when:

- A new employee is recruited
- An employee is transferred from one department to another
- Changes in process requirements
- A new process demands a special skill set
- A job position is changed

Aside from this, even if nothing has changed, a biannual audit should be conducted to ensure that everything is accurate. Thus, skill management is an integral part of a visual organization.

After every stage in skill development (stages 0 to 4), the established learning goals must be satisfied. Each learning stage should have a learning outcome that should be verified. Use verbs such as describe, demonstrate, analyze, list, and others, to ensure that the student can perform the task as well as intended and finally can train others in that task.

ON-THE-JOB TRAINING

In today's manufacturing- and service-providing organizations, tools, software, and technologies are changing rapidly. Skill development becomes urgent. In such circumstances, OJT becomes a very common form of training for new and experienced employees.

There are three main aspects of OJT:

- Prepare the trainee
- Prepare the training material and the trainer
- Evaluate the training delivered and the lessons learned

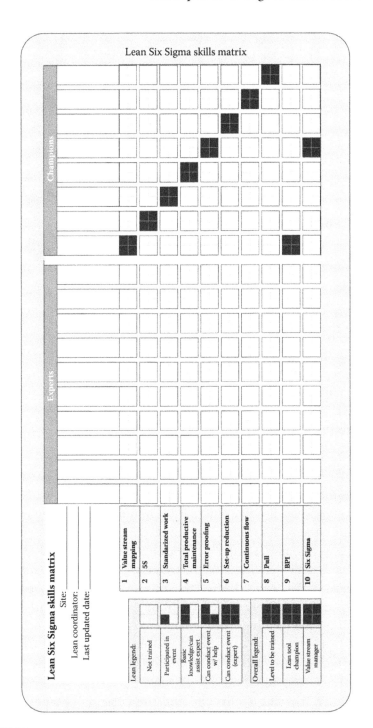

FIGURE 4.2
Lean Six Sigma skill matrix.

Level	Description	Explanation	
0	Can't do the job	Insufficient knowledge and experience to perform standard work	
1	Knows all steps of the task	Has fully understood documentation, tools and is familiar with the tools and the job	
2	Can do the job	Has received instructions from instructor Has performed task correctly in front of the instructor	
3	Can do fully	Has been certified by the instructor	
4	Can teach others how to do	Has taught and audited another person's work within 90 days	

FIGURE 4.3
Skill assessment level guide.

Prepare the Trainee

The person undergoing the training must be prepared prior to the training. In other words, this person should be willing to learn, has the required education/ability to absorb and understand the training material, and is motivated. This is normally ascertained by the human resource function before recruiting a person.

Prepare the Training Materials and the Trainer

A good trainer must know the job and be able to teach the trainee. Normally, the most experienced employee is selected to train the new employee. This can misfire because this trainer may take certain job aspects for granted or treat them as trivial—leaving the trainee in the dark. In other words, the trainer needs to be well-versed in the training skills. The trainer has to be able to transfer the knowledge and the intricacies of the job to the trainee so that the trainee can become effective and do the job efficiently.

Facilitating the training is a human resources/training function. The training materials are first prepared by the appropriate departments—engineering, systems, maintenance, finance, etc., as appropriate.

Evaluate the Training Delivered and the Lessons Learned

After the OJT has been completed, it is necessary to evaluate what went well, what did not go well, and what changes (if any) need to be made. The findings should be documented. The plan–do–check–act (PDCA) cycle for continual improvement in the OJT should be followed. OJT is very economical because the training is in-house and it is imparted by experienced employees. The trainee gets immediate feedback from the manager.

COACHING AND MENTORING

We expect greatness from all our people. We expect them to accept and conquer challenges that may seem overwhelming at first sight. The greatness in people comes out only when they are led by great leaders. We are all growing and learning and we all need teachers and coaches to help guide us. We say at Toyota that every leader is a teacher developing the next generation of leaders. This is their most important job.

Akio Toyoda
President of the Toyota Motor Corporation

The managers and leaders of an organization who are determined to understand and adopt Lean philosophy have their task defined by the Toyota chief—improve the capabilities of the people around you.

Job description of a coach if such a position needs to be created:

- Performing skill assessments to find out what skills your department has and which skills are needed. Determine the gaps.
- Training through OJT and cross-training.
- Train employees to follow standardized work.
- Seek opportunities for improvements through kaizen activities.
- Promoting teamwork by stepping in and helping the team to overcome a problem.
- Develop potential leaders. Find them and give them opportunities to solve problems by finding such opportunities.
- Share, spread, and communicate information arising out of lessons learned for the benefit of all people.
- Monitor and help to implement ideas from suggestion schemes. Use these opportunities to see how the team members think. What is their thought process? Nurture their ideas and provide them a safe and secure (without the fear of losing their jobs) environment.
- Problem-solving capability.

- Finally, a coach is a lifelong learner learning new things. By preparing to teach and observing human nature, the coach attains more skills and becomes a better teacher.

Mentoring is a long-term commitment between the mentor and mentee to work together. Mentors teach and guide a mentee. The mentee works on the issue to learn and grow. In the process, the mentor and the mentee reach a stage where "I don't know" is an acceptable and valid answer, and when "I don't know" is the answer, it is time to go and see! (Rother 2010)

Even top executives should seek out mentors to help them become better at what they do. Lifelong learning is important to those in leadership positions.

Here are some practical suggestions for making people more creative.

Years of experience have shown me that almost everyone is far more creative than he or she is ever allowed to be at work. The talent is sitting there—setting it to work can be exciting. Here's how:

- Telling people to be creative provokes an opposite response. On the other hand, praising every trace of creativity you find—in someone's suggestion or even something another company has done—prompts people to believe that creativity is what's wanted, admired, and rewarded. That alone can be quite stimulating.
- Using brainstorming is more difficult than it looks. Make it clear that people must come equipped with prepared ideas—and the more the merrier—you will have a richer field to work with.
- Many people have their best ideas on the way home, just before or after sleeping or, in the shower. It's well understood by neuroscientists that often the best way to solve a problem is to look away from it. Discuss this with your teams and make it clear that taking breaks and disrupting routines can be productive. I find it especially helpful to ask myself a question before I go to sleep.

TEAM MANAGEMENT

Obviously, teams can outperform individuals. Teams offer more than just increased efficiency. Because the team members are empowered to deal with many things that affect their work, teams provide a great source of job satisfaction and employee involvement.

A team can be beneficial when:

- A complex and companywide task like Lean implementation must be addressed
- Creativity is required
- The path forward is not clear
- Improved efficiency of resources is required
- Learning needs to be fast
- A high level of commitment is desired
- Cooperation is required to implement the plan
- The task or the process is cross-functional

When traditional methods of problem solving fail, and decision making or fast implementation like Lean methodology is required, teams can help meet these challenges. Teams are not an answer to every conceivable problem. However, if properly formed and well conceived, their potential is vast.

For example:

- Teams can reduce lead times
- Decrease cycle times
- Operate business units
- Redesign products and systems
- Identify customer needs

The list can go on and on…

The types of teams can be defined by their objectives.

- A functional team is composed of a manager and the employees in the department. Here, issues like decision making, objectives, and leadership are simple and clear.
- The problem-solving team offers ideas and suggestions on how processes can be improved. Regular meetings are held.
- A fully self-managed team selects its own members. Members evaluate each other's performance. Here, the supervisor's position has less importance and can also be eliminated.
- A cross-functional team has members from various disciplines but of the same hierarchical level. Complex issues and problems can be tackled by this team.
- A virtual team uses computers and information technology to tie together physically separated members to achieve a common goal.

TEAM DYNAMICS

Team formation takes time. Teams go through the five famous stages. These are the forming, storming, norming, performing, and closing/adjourning stages (Figures 4.4 and 4.5).

Forming

Characteristics:
- Positive expectations
- Moderate eagerness to get started
- Low accomplishment

Feelings:
- Excitement, anticipation, and optimism
- Initial tentative attachment to the team
- Suspicion, fear, and anxiety about the job ahead

Behaviors:
- Attempts to define tasks and how they should be accomplished
- Attempts to define group behavior and how to deal with group problems
- Complaints about the organization and barriers to the task

To move beyond the forming stage:
- The leader sets the direction
- The group defines and develops its purpose
- Team members begin problem solving and decision making

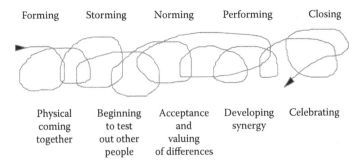

Teams go through phases

Forming	Storming	Norming	Performing	Closing

| Physical coming together | Beginning to test out other people | Acceptance and valuing of differences | Developing synergy | Celebrating |

FIGURE 4.4

Process of development through which teams pass.

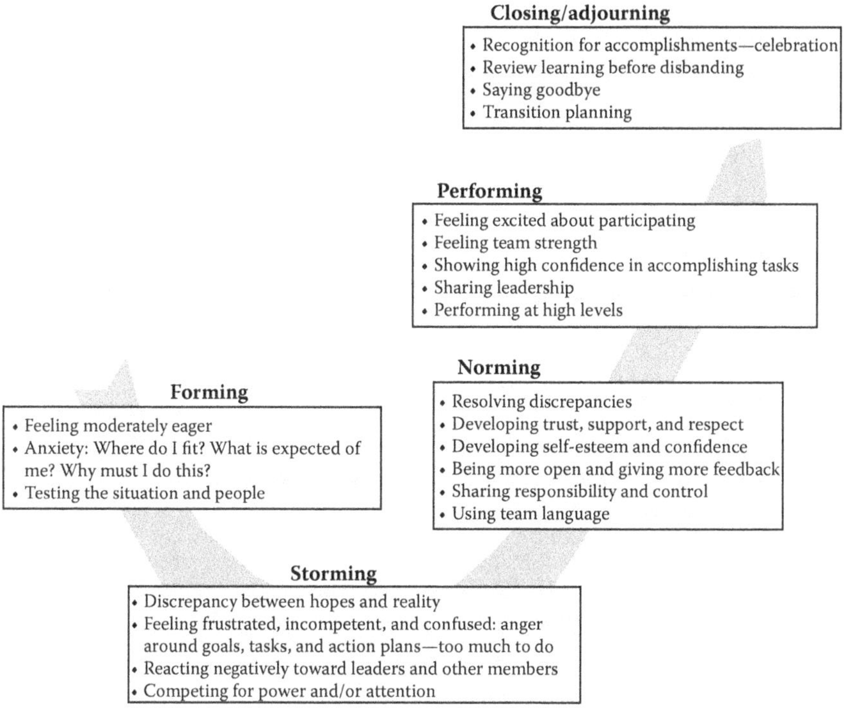

Closing/adjourning
- Recognition for accomplishments—celebration
- Review learning before disbanding
- Saying goodbye
- Transition planning

Performing
- Feeling excited about participating
- Feeling team strength
- Showing high confidence in accomplishing tasks
- Sharing leadership
- Performing at high levels

Forming
- Feeling moderately eager
- Anxiety: Where do I fit? What is expected of me? Why must I do this?
- Testing the situation and people

Norming
- Resolving discrepancies
- Developing trust, support, and respect
- Developing self-esteem and confidence
- Being more open and giving more feedback
- Sharing responsibility and control
- Using team language

Storming
- Discrepancy between hopes and reality
- Feeling frustrated, incompetent, and confused: anger around goals, tasks, and action plans—too much to do
- Reacting negatively toward leaders and other members
- Competing for power and/or attention

FIGURE 4.5
Stages of team development.

Storming

Characteristics:
- Goals and structure are clear
- Group skills are gradually increasing
- Accomplishments of tasks is increasing slowly
- Motivation plummets because there is a discrepancy between the initial hopes and reality. Team becomes dissatisfied with its discrepancy with authority

Feelings:
- Resistance to task
- Changes in attitude about the team and chances for success

Behaviors:
- Arguing among the members
- Defensiveness and competition
- Questioning team members

To move beyond the forming stage:
- The team must begin to share leadership tasks
- Define roles, responsibilities, policies, and procedures
- Share philosophies with other team members
- Be supportive of one another
- Create system necessary to support the team action

Norming

Characteristics:
- Productivity continues to rise
- Expectations fall in line with reality
- Satisfaction increases dramatically because the structure and progress are clear and relationships are built and have become settled

Feelings:
- A new ability to express constructive feedback
- Acceptance of membership in the team
- Relief that everything will work out

Behaviors:
- An attempt to achieve harmony by avoiding conflict
- More friendliness in trusting and sharing of personal problems
- A sense of team unity and a common spirit and goals
- Establishing and maintaining team ground rules

To move beyond the forming stage:
- Acknowledge the progress of the group
- Make sure that the team members are empowered

Performing

Characteristics:
- The group is positive and eager
- Group productivity, pride and confidence are high
- Members work autonomously
- Members go out of their way to help each other

Feelings:
- Insights into each other's strengths and weaknesses
- Satisfaction in the team's progress

Behaviors:
- Constructive self-challenge and improvement
- Ability to prevent or work through the group problems
- Close attachment to the team

To move beyond the forming stage:
- Keep communications open
- Focus energy on results
- Recognize and support member's competence and accomplishments
- Help the group execute, follow through, and follow up
- Continue to conduct problem solving sessions to keep morale high

Closing/Adjourning

Characteristics:
- The team's mission is complete
- Improvements have been or are being implemented
- A record has been made of the team's progress and conclusions
- A final report has been submitted

Feelings:
- Nostalgia, reminiscence, and mourning about what it was like working with the team in the beginning versus more recently and what was learned in the experience
- Accomplishment
- Loss and ending

Behaviors:
- Promises to stay in touch
- Have regular reunions

To bring closure:
- Celebrate the team's successes
- Acknowledge and allow others to acknowledge your accomplishments
- Say goodbye

HANDLING PROBLEM PEOPLE

First, we should not:

- Be defensive and argue with them, preach, lecture, or threaten them
- Ignore them
- Criticize them
- Tell them where to go
- Put them down by ridicule or shame

Some ways for handling problem individuals include:

- Ignore their "wisecracks" or jokes
- Inform the group when you feel that a subject lies outside the scope of the training
- Face the problem squarely and state the reasons as you know them
- Defer the problem to a later date—give yourself time to think about it or talk it over with others
- Defer the problem to a private meeting during a break or after the session with the person who raised it
- Don't be afraid of parting ways if a person is still problematic

CONFLICTS

Conflicts are roadblocks to progress. Conflicts can be resolved. Some effective steps include:

- Identify the interests of each person by asking "What do you want?" Then, listen to the answer you get. People's interests are issues that cause conflicts. We usually only know their positions, which is probably obvious.
- Identify higher levels of interest by asking, "What does having that do for you?" It's important to understand what people really want.
- Create an agreement frame by asking, "If I show you how to get X (X is person's real interest), would you do Y (Y is what you want from the person X)?"

- Brainstorm for solutions together to find a win–win solution. Do not just give a solution and expect the other person to accept. A win–win solution must satisfy interests of both parties. You get commitment by getting people involved.

TEAM DECISION AND CONSENSUS BUILDING

When a decision is reached by consensus, each team member should be able to honestly say:

- "I believe the team understands my point of view"
- "I believe I understand the points of view in the team"
- "This might not have been my first choice, but I can live with it"
- "I will support this decision because it was reached in an open and fair manner"

Whenever possible, use consensus building to optimize communication, motivate team members to work together on implementation, and minimize the risk of sabotage by disgruntled members.

Reaching Consensus: Some Tips

- First, agree on what consensus means.
- Set a time limit for making the decision—and a fallback if no decision is reached.
- Check to see how much consensus already exists (write proposal on a flip chart). Discuss only the point of concern.

Another way to check for existing consensus is to conduct a straw vote.

Each member writes a number from 1 to 5 on an index card to indicate readiness to endorse the decision. Write "5" for complete agreement and "1" to strongly disagree. If the team gets stuck, break the team into smaller groups to reach consensus. Smaller teams present views to the larger team.

One key rule: If a participant disagrees with a decision, he or she must explain why and offer an alternative to help avoid a stalemate.

Finally, try to select a team member with some or all of the following traits:

- Has a balance of "hard" and "soft" skills
- Best experience possible

- Has knowledge of the subject
- Has willingness to join
- Is available
- Is a good listener
- Can give and take feedback
- Can communicate clearly
- Is mature enough to take responsibility
- Has good follow-through on commitments
- Has leadership and managerial skills

SUGGESTION SCHEMES

Employee suggestion schemes have been in existence for over a century. It can be in the form of a simple box on the wall or a fully developed system led by the managers and coaches.

The box on the wall system, in which employees hand over their ideas to management, has its own problems. Many ideas are submitted but the management does not have enough resources to handle them. This results in frustration among employees because their suggestions are not implemented.

The Lean suggestion system is where:

- All ideas are accepted
- An appreciation and recognition is given for the suggestion
- Coaches help to grow the idea
- The originator of the suggestion implements his/her idea

Steps to make the suggestion system successful:

- Get as many people involved as possible
- Focus on small ideas and small successes
- Don't micromanage ideas
- Keep them visible
- Keep the recognition plan simple—usually immediate small payments/ prizes

Organize all ideas/suggestions received using a simple idea board. Most importantly, realize that people want to be recognized even if they don't show it.

THE IDEA BOARD

There are many other types of idea boards. The idea is to make it easy to prioritize and follow up the ideas until their successful resolution, implementation, and subsequent control.

The real value of the suggestion system lies in its simplicity, which is designed for the employees. The system is free of all complicated form fillings, and cost–benefit jargons. These are simple ideas to create a Lean mindset that everyone can contribute to the overall growth of the company through their ideas (Figure 4.6).

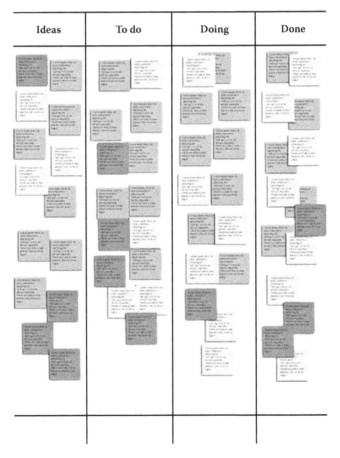

FIGURE 4.6
Idea board.

SAFE WORKING ENVIRONMENT

The work environment safety plays an important role in improving working conditions, employee safety, and productivity.

Under government safety laws, it has been emphasized that workers are entitled to working conditions that do not pose a risk or hazard of serious physical harm. Hazards can range from repetitive motion issues to life-threatening hazards.

Poor ergonomics has a very big impact on employee safety. Ergonomic injuries account for more than 50% of all workplace injuries. The most important ergonomics risk factors are posture, force, and repetition. Understanding these factors will enable the organization to reduce waste by eliminating non–value added work. Additionally, improving safety will reduce insurance costs, lost time, and regulatory penalties, etc.

Section II

Principles of Continuous Improvement Process

5

Importance of Principles, Systems, and Tools in Continuous Improvement

The continuous improvement (CI) process begins by clearly defining value in the eyes of customers, both internal and external. Expectations of customers and stakeholders must be clearly defined and unambiguously communicated so that the CI process can be designed to meet customer needs.

Dr. Shigeo Shingo said (Model and Application Guidelines 2010–2011), "Improvement means the elimination of waste, and the most essential precondition for improvement is the proper pursuit of goals. We must not be mistaken, first of all, what improvement means."

The four goals of improvement must be to make things

- Easier
- Better
- Faster
- Cheaper

New Lean practitioners may be confused with the Shingo approach because Lean is generally taught with a focus on tools and results. Focus on tools and results leads the Lean practitioner away from the CI philosophy preached by Shingo—a time-tested approach that has made Toyota a world leader. Figure 5.1 illustrates the connection between CI principles, CI systems, and CI tools.

In a typical Lean implementation attempt, organizations start on the left-hand side, providing training on tools to Lean teams to make improvements. They build systems and believe that people will eventually discover the principles behind the tools they are using. The intention is not to say that the focus on tools, teams, etc., is wrong, but I do believe it is not complete. I think we also need to look at working from right to left—to help people

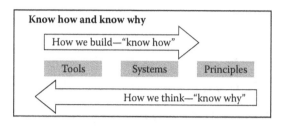

FIGURE 5.1
CI principles, systems, and tools to know why and then to know how.

understand the guiding principles, to think about the kinds of systems they want, and to use tools to design and redesign those systems. Dr. Shingo said, "People need to know more than how, they need to know why."

The 10 guiding principles in Figure 3.2 are like the laws of science. They do not need proof, they are eternal, and they govern consequences for everyone. It is immaterial whether one believes them or not. Think in terms of Newton's law of gravity. You may or may not agree in gravity, but it governs you regardless of what you believe. The same is true regarding any of the guiding principles, such as "respect for every individual," "focus on process," or "constancy of purpose." You may not agree with them (or may not even be aware of them), but they govern you.

WHAT IS A PRINCIPLE?

Principles Govern Consequences

Principles are not the same as values. Many organizations have their values listed everywhere in their organization. Many will provide all employees with a list of values that they can keep in their wallet or on their employee name badge.

Values are social norms. They are personal, emotional, subjective, and arguable, and they are not the same as principles (Figure 5.2).

Now see Figure 5.3 to understand the difference between values and principle and how the principle can govern the consequence. It shows a cartoon where a lady has been mugged and a policeman is taking her statement. She says, "They were your height, but thinner and better looking. You haven't caught them yet, so obviously they're smarter."

If you looked at the policemen's badge, you might find the values of "teamwork, loyalty, precision, and innovation" on the badge. You can find

Think "gravity"...

whether you believe
in gravity or not, there
are consequences for
understanding it (or
not understanding it)

FIGURE 5.2
Principles that govern consequences.

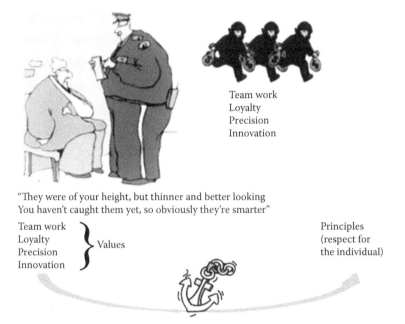

Team work
Loyalty
Precision
Innovation

"They were of your height, but thinner and better looking
You haven't caught them yet, so obviously they're smarter"

Team work
Loyalty } Values
Precision
Innovation

Principles
(respect for
the individual)

FIGURE 5.3
A cartoon of a mugged lady and a policeman.

the same with the thieves. They could have the same values on their ID cards. They could work well as a team, be loyal to each other (not to be an informer against the others), be very precise in their work, and quite innovative in the way their carry out their work (steal).

The two groups may have the same values, but one group (policeman) may have those values anchored to a guiding principle, that is, "respect for every individual." This makes all the difference. Thus, seek out the underlying principle and how the values need to be anchored to the guiding principles to have a clear understanding of the ideal behavior you want to see in your organization. See more examples of how values are anchored in principles in Table 5.1.

Values, when anchored with principles, lead to ideal behaviors. If an organization having values such as team work wants to understand the distinction between principles and how their values need to be anchored to guiding principles, they must ask people to pick a value ("teamwork") and to describe what this would look like in terms of behaviors that they could see. In this case, the behavior responses to the value "teamwork" could be as follows:

- Respect every individual
- Lead with humility
- Create value for the customer

In summary, Lean leadership must learn, practice, and follow the 10 Deming principles:

1. Respect every individual
2. Lead with humility
3. Focus on the process
4. Embrace scientific thinking
5. Follow and pull value
6. Assure quality at the source
7. Seek perfection
8. Create constancy of purpose
9. Think systematically
10. Create value for the customer

These principles not only govern the consequences but also create a vision and strategy for organizational excellence.

TABLE 5.1

How Principles Determine Values

Value	Shingo Guiding Principle
Make our customers the focus of everything we do	Create value for customer
Recognize our people as our greatest asset	Respect every individual
Treat each other with respect	Respect every individual
Strive for excellence	Seek perfection
Keep our commitment	Constancy of purpose

6

Principles of Continuous Improvement Process

"The Principles of Continuous Improvement" chapter includes seven sections:

1. Process focus (with five subsections)
    - Focus on the process
    - Spend time on the floor
    - Look at the process in actual operation
    - Measure the process to catch disruptions and problems
    - Teach the importance of process focus
2. Identification and elimination of barriers to flow including waste (with seven subsections)
    - F.W. Taylor economics and flow
    - Functional silos
    - Quality as a specialized function
    - Cost accounting
    - Nine wastes (Muda)
    - Mura and Muri
    - Not managing the flow
3. Match the rate of production to the level of customer demand (with two subsections)
4. Scientific thinking
    - MBF (management by fact)
    - Go and see
5. Jidoka (autonomation—combination of automation and mistake proofing)
6. Integrate improvement with work
7. Seek perfection

PROCESS FOCUS

Processes in general are administered by the supervisor and the team leader under the regular watch of the process manager. For reference, review the Toyota Supervisor and Toyota Team Leader job description and specification in the appendix.

- Focus on the Process

    Team leaders, supervisors, and managers play a major role in "how to focus on the process." For them, it includes spending as much time as possible "differently" and learning to look for, ask about, measure, and teach different things.

- Spend Time on the Floor

    The first thing Lean advises: go to the gemba (gemba—where the action takes place—the shop floor/activity center) and spend most of your time (80%) if you are a team leader, at least once a week (preferably one hour) if you are an operations executive, and proportionately for other positions in between. Go to the place when the process is operating. Talk with people doing the work; observe the process as it is being performed. The idea is to anchor yourself in the regular, fresh, and new information about the reality of the process. Being there is the only way you can do this.

- Look at the Process in Actual Operation

    To be most useful, measure what you see with a standard or with expected performance. Is the practice you see as per the standardized work? Learn what people really deal with daily. When actual outcome fails to meet expected outcome, ask people about it. What happened? How often does it occur? Where is it documented? Can I look at it? Is problem-solving under way for this issue?

- Measure the Process to Catch Disruptions and Problems

    Normally, the operations have many measures such as quality defects, productivity, on-time delivery, scrap, etc., which are the lagging measures. Much water has flown down the river by the time actions are taken on these measures. Process measure consists of measuring the actual process performance when the process is running against the expected performance. For repetitive manufacture, the frequency of the measure may be every takt time, as often as

10, 12, or 15 minutes. For a drawing office, it may be once or twice a day; for order processing, the frequency may be four times a day. For all changeovers, the frequency will be every changeover and its "handoff" points.

The purpose of process measure is to identify, isolate, and highlight process misses and process disruptions. This distinguishes the Lean mind-set from the conventional world. Conventionally, we are conditioned to work around problems—not to face the problems but to keep them on one side and carry on to meet the schedule.

Contrarily, a Lean thinker devises measures to highlight problems even if a temporary work-around is necessary. The Lean thinker looks for the root cause and then eliminates it. This proves the point that Lean is an improvement system.

- Teach the Importance of Process Focus

When you are in the gemba, people ask questions and ask for responses. Your response should be explicit. "You identify production problems that slow you down or interrupt you, and we will eliminate them or give you a good reason why not." When you keep your end of the bargain, you have successfully taught the importance of the process focus.

Through timely shop floor visits, leaders and floor level people become more engaged with each other. With further improvements and clues about the process, you experience a steady stream of favorable changes that collectively make a considerable difference in performance. In this way, process focus is an important key to unlocking the "gold mine" of continuous improvement.

IDENTIFICATION AND ELIMINATION OF BARRIERS TO FLOW

The following are the barriers to flow (flow of a product/service provision value stream):

- F.W. Taylor Economics and Flow
- Functional silos

- Quality as a specialized function
- Cost Accounting
- Nine Wastes (Muda),
- Mura and Muri
- Not managing the flow

F.W. Taylor Economics and Flow is one item which is a barrier to flow! So please treat this item as any other—like Functional Silos.

F.W. Taylor's (1856–1915) scientific management technique has been employed to increase productivity and efficiency both in private and public sectors. It has also had the disadvantages of ignoring many of the human aspects of employment. This led to the creation of boring repetitive jobs with the introduction of systems for tight control and the alienation of shop floor employees from their managers. The operators are asked to come to work with their hands only, leaving their brains at home! This division of labor as a central concept of the "best way" is still prevalent in many organizations.

Functional Silos

When you think of a silo, you think of the type found on farms, full of grain (Figure 6.1). Full grain, full harvest, full happiness—it sounds like a perfect metaphor for a successful organization.

The silo effect in a business refers to the lack of communication and cross-departmental support. Each function behaves like "silos." Functional teams work only on their own goals, often ignoring the needs of others, and information (and customers) gets lost in the middle. The cycle repeats—ignoring company-wide objectives and external customer expectations.

This results in a dangerous downward spiral resulting from the "silo" mind-set.

Each division is focused on its own primary objectives

Divisions do not communicate sufficiently or effectively because of the factors that feed into silo formation

Divisions do not cooperate with one another because they are uninformed

The lack of cooperation breeds resentment and a reciprocal lack of giving

Division members feel isolated from other divisions and focus work effort within their own divisions

Each division focuses on its primary objectives

Quality as a Specialized Function

Whenever a problem arises, you hear the supervisors and departmental managers cry out, "call the quality control guys" or "call the Six Sigma guys." The problem and its solution often gets delayed or lost in endless

FIGURE 6.1
Silo towers.

meetings and discussions. This practice hides the problems and opportunities from management and creates a demoralizing atmosphere for workers.

This specialization is most damaging because such functions "design" the work in an office away from the work area.

Cost Accounting

Cost accounting believes that a process is equal to the sum of its operations and cost can be improved by improving the individual operations. In a Lean system, the whole is not equal to the sum of its parts because an individual operation could be controlled to the last second but the parts or a product can sit in queues or a warehouse for months with no further advancement toward the customer or toward getting paid.

Nine Wastes (Muda)

Muda is a Japanese word for waste or any work that the customer is not willing to pay for. This is opposite of the value. We can say that all work consists of value-added, non–value-added, and non–value-added but essential work.

The American Society for Quality (ASQ) defines non–value-added as a "term that describes a process step or function that is not required for direct achievement of process output. This step or function is identified and examined for potential elimination."

This represents a shift for production and manufacturing engineering, which have traditionally studied ways to improve value-added functions and activities (e.g., how this process can run faster and more precisely).

Lean thinking does not ignore value-added activities but does shine a spotlight on waste (muda, mura, and muri; Figure 6.2; Table 6.1).

Muri (Overburden, Stress)

Muri is the Lean waste that occurs when a person, machine, or process is pushed beyond the normal or designed limits (Figure 6.3). For persons, this means pushing the body at a working rate that could cause fatigue, injury, or defective work. Working in an unsafe environment could also cause injuries and accidents. Forcing a machine to work beyond its designed limits results in a breakdown or causes repeated corrective maintenance calls. Process, when it is pushed to the limits, can cause the data handling to go haywire!

9 Wastes

Overproduction

Producing more, sooner or faster than the next job requires
- Building more than the customer wants?
- Where is the additional parts/paperwork stored?
- How much did this cost?

People movement

- How much excess movement do you do each day?
- What else could you have done with the time?
- How widely spaced are your work centres?

Material transport

- How many times are materials handled?
- Multiple sign offs?
- Are your schedules based on full loads and efficient routing?

Rework and defects

- Scrap
- Yield
- Wastage
- Invoice errors, order errors
- How much does a correction cost?

Inventory

Stock and work-in-progress
- How much does this cost?
- Why does it happen?
- How much space is taken up by stock?

Inappropriate processing

Procedures and paperwork
- (Anything fiddly) Could things be done more simply?
- Do you re-enter data?
- Is paperwork fit for purpose?
- Producing a few extra?

Talent

- Have you realized the full potential of your workforce?
- What extra skills do people have?
- How could these be used?
- What will be the benefit?

Delays and waiting

Waiting for:
- Paperwork
- Instruction
- Supervision
- Products
- Suppliers
- Customers

Need some help tackling the 9 Wastes?
Contact Manufacturing Advisory Service in the South West
0845 608 3838
info@swmas.co.uk
www.mas.bis.gov.uk/south-west

MAS is the trusted provider of advice and hands-on support for manufacturers aiming to achieve sustainable success.

Energy

- Do you switch off lights, printers and machines when not in use?
- Is the roof insulated and are the sky lights clear?
- Are minor repairs made immediately?

If you would like this information in another format please call us on 0845 608 3838.

Commissioned by

FIGURE 6.2

Nine wastes. (Courtesy of the Department of Business Information.)

TABLE 6.1

Nine Wastes (Muda) and Their Definitions, Characteristics, and Causes

Waste	Definition	Characteristics	Causes
Overproduction	• Producing more than needed • Producing faster than needed	• Accumulated inventory • Extra equipment/oversized equipment • Unbalanced material flow • Extra part storage racks • Extra manpower • Batch processing • Excessive capacity/investment • Additional floor space/outside storage • Large lot sizes • Building ahead	• Incapable processes • Just-in-case reward system • Lack of communication • Local optimization • Automation in the wrong places • Cost accounting practices • Low uptimes • Lack of stable/consistent schedules
Waiting time	• Idle time is produced when two dependent variables are not fully synchronized • Man wait time • Machine wait time	• Man waiting for machine • Machine/materials waiting for man • Unbalanced operations (work) • Lack of operator concern for equipment breakdowns • Unplanned equipment downtime	• Inconsistent work methods • Long machine changeover times • Low man/machine effectiveness • Lack of proper equipment/materials

(Continued)

TABLE 6.1 (CONTINUED)

Nine Wastes (Muda) and Their Definitions, Characteristics, and Causes

Waste	Definition	Characteristics	Causes
Unnecessary transport time	• Any material movement that does not directly support a Lean manufacturing system	• Extra carts, forklifts, dollies • Multiple storage locations • Extra material racks • Extra facility space • Incorrect inventory counts • Damaged material	• Large lot processing • Unleveled schedules • Lack of 5S • Lack of visual controls • Improper facility layout • Large buffers and in-process kanbans
Overprocessing	• Effort which adds no value to a product or service • Enhancements which are transparent to the customers or work which could be combined with another process	• Process bottlenecks • Lack of clear customer specifications • Endless refinement • Redundant approvals • Extra copies/excessive information	• Engineering changes without processing changes • Decision making at inappropriate levels • Inefficient policies and procedures • Lack of customer input concerning requirements
Inventory build-up	• Any supply in excess of process requirements necessary to produce goods or services just in time	• Extra space on receiving docks • Material between processes • Stagnated material flow • LIFO instead of FIFO • Extensive rework when problems surface • Long lead time for engineering changes • Additional material handling resources (men, equipment, racks, and storage space)	• Incapable processes • Uncontrolled bottleneck processes • Incapable suppliers • Long changeover times • Management decisions • Local optimization • Inaccurate forecasting systems

(Continued)

TABLE 6.1 (CONTINUED)

Nine Wastes (Muda) and Their Definitions, Characteristics, and Causes

Waste	Definition	Characteristics	Causes
Wasted movement	• Any movement of people that does not contribute added value to the product or service	• Looking for tools • Excessive reaching or bending • Materials too far apart (walk time) • Equipment for moving parts • Extra "busy" movements while waiting	• Equipment, office, and plant layout • Lack of 5S • Lack of visual controls • Inconsistent work methods (standardized work) • Large batch sizes
Rework and making defective product	• Repair of a product or service to fulfill customer requirements	• Extra floor space/tools/equipment • Extra manpower to inspect/rework/repair • Stockpiling inventory • Complex material flow • Questionable quality • Missed shipments/deliveries • Lower profits due to scrap • Reactive organization	• Incapable processes • Excessive variation • Incapable suppliers • Management decisions • Insufficient training • Inadequate tools/equipment • Poor layouts/unnecessary handling • High inventory levels
Unused talent	• Nonuse of full potential of the employees	• Indifferent attitude of employees	• Lack of skill development programs. Incomplete Lean transformation
Energy waste	• Wasted electricity/gas/water/compressed air	• Doors kept open, roof not insulated, skylights absent or dirty, light/motors not switched off when not in use, repairs and maintenance not done in time	• Lack of systematic energy conservation programs. ISO 14001 not implemented

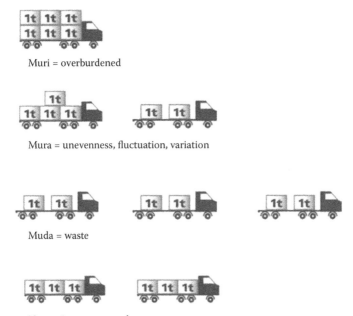

Muri = overburdened

Mura = unevenness, fluctuation, variation

Muda = waste

No muri, mura, or muda

FIGURE 6.3
Muri, mura, and muda explained.

In the factory, an assembler has three documents—a fixture instruction, an assembly drawing, and a bill of material—and none of them agree. What should she do? This is an example of personal (mental) muri, which, in the worst cases, can cause not only migraine but can also lead to a serious human resources problem.

Mura

Mura is variation, fluctuation, unevenness, or deviation. Mura results from poor and unplanned schedules, unplanned changeovers, and lack of standards. The worst form of mura is tampering or overadjusting a process to control it without understanding the process inputs and output relationships. This causes additional variations resulting in the product or service to go beyond the specifications that are asked for by the customer.

The effects of mura can be managed by leveling the production, variety reduction, and just-in-time using a pull system or a continuous flow (one piece flow) method.

Not Managing the Flow

Unmanaged and invisible flow is the biggest barrier to flow. To manage the flow, control the flow and make it visible.

The first step to managing the flow is to prepare the standard work. Standard work makes the process less wasteful by

- Removing non–value-added activities from the process steps
- Connecting and aligning the process steps to maximize customer value
- Aligned steps complete the process, which is carried out according to a customer's demand (takt time)

The next step is to control the flow and make it visible from end to end for all persons involved in the process. This is accomplished through value stream mapping. This is a powerful graphical technique to document and communicate the current and ideal future state of the value stream. The stream is an unhindered flow of processes. "Value" is an efficient and effective process.

Value stream, in the Lean vocabulary, describes a business process. A business process providing value to the customer can be any one of the following and more:

- Quote-to-cash
- Procure-to-pay
- New product or service development
- Day-to-day order fulfillment

MATCHING THE RATE OF PRODUCTION TO THE LEVEL OF CUSTOMER DEMAND

Just-in-time is the ultimate goal of Lean implementation. This means the evolution of a process to the level in which it is able to produce exactly what the customer wants, when the customer wants it, and in the right quantity. Just-in-time is completely opposite to overproduction—overproduction is the *worst* of all nine wastes.

Just-in-time is supported by continuous flow, kanban or pull system, and heijunka or production leveling.

SCIENTIFIC THINKING

Scientific Thinking Process

- Find and define the problem
- Consider the problem as an opportunity to improve
- Observe the process, record the data and information, and "measure" the process
- Determine and define the proposed solution
- Validate, test, and carry out a pilot run of the proposed solution
- Compare results before and after the task
- Document the new process steps; ensure all affected people are trained
- Go back to step 1 and continue the plan–do–check–act cycle

Management by Fact

Decisions for improvements must be based on data. This is called management by fact or management by data. The form and example are shown in Figure 6.4.

Go and See

Gut feeling, perceptions, incomplete standards, and past experiences are used to make decisions—usually by sitting in an office or a meeting room. This can lead to changes that can make the process difficult to control and create more waste. By directly watching, true and accurate information and data can be gathered and confirmed.

JIDOKA (AUTOMATION—COMBINATION OF AUTOMATION AND MISTAKE PROOFING)

Respect for people is at the very root of the word jidoka (Table 6.2). As you can see, jidoka and just-in-time are the two main pillars of Lean. Jidoka is the more important pillar out of the two (Figure 6.5).

MBF Example (Scrap reduction Project)

OBJECTIVES AND PERFORMANCE TRENDS	PRIORITIZATION & ROOT CAUSES	COUNTERMEASURES/ ACTIVITIES	ASSESSMENT OF PROCESS CAPABILITIES

Problem Description:
high level of scrap

Objective:
to reduce the scrap level 1.39% to 80% Aug.

Performance Metric:
Scrap vs. Sales cost (%)

% CONTRIBUTION

Scrap vs. Sales cost (%)

ACTIVITIES

WHAT	WHO	WHEN
Main contributors analysis	R. M.	5-Jul-02
		done
Handling material improvement to avoid broken wire and lead wire damage	M. Ch.	31-Aug-02
Scrap metric review (Incomplete coil clasification)	R. M. E.C.	23-Aug-02
Pilot run to evaluate a re-design on 827 model to improve O.C.V.	R. E	31-Aug-02
To evaluate the klixon insulation system using a tube mylar 806 and 827	O. G. R.M.	8/31/2002

CORRECTIVE ACTIONS

WHAT	TYPE	WHO	WHEN	IMPACT
E.C.R. to use lead wire instead wire tin in klixon (jumper)	1	E. C.	15-Jun-02	25%
To change Press arms (7.50" to 7.70" wide dimension) 806 and 827	1	M. Hdz.	15-Jul-02	15%
New tester on lamination area	2	J.C. Herrera	29-Aug-02	12%
ECR new design 827	1	R. Elizondo	Sep.- 26-02	30%

Comments:

Contributors code

FH	Broken wire on Transformer
WH	Incomplete coil
FQ	Hi-pot
FR	High excitation
WE	Fail out wire
G1	Low O.C.V.
FF	Scrap from Ballast repairing
FJ	Damage coils
LJ	Scrap from coils repairing
W1	Clean wire process

Transformer bad handling

Scrap metric

Winding machine rack system

Product design

Low stack at press operation

winding machine set-up

Target .50%

Target .80%

37316	37346	37379	37409

Background:

Currently we have high scrap level on Slim Line Area (1.39% avg. according with Monthly data of March–June 02), impacting on manufacturing's costs

Savings: Reduce scrap to lesss than 0.8%

Start Date: Jul-02

End Date: Sep-02

Blackbelt: Romualdo Maldonado

Team: M. Chavira, R. Elizondo, M.Hernandez, M. Pañola, J.Cadena

FIGURE 6.4
Management by fact form.

TABLE 6.2

Jidoka and Just-in-Time, the Two Main Pillars of Lean

Goals of Toyota Production System	
Highest Quality, Lowest Cost, and Shortest Lead Time	
Two Pillars of the Toyota Production System	
Jidoka	**Just-in-Time**
Build quality in	Takt time
• Stop the line	• Standardized work
• Find abnormalities	• Labor saving
Flexible manpower	Continuous flow
• Separate human and machine work	• One piece flow
	• Synchronization
	• Multiprocess handling
	• Equipment deployment
	• Streamline
	• Small lot size
	Pull system
	• Kanban tool to transport tools and information
Base of Toyota Production System	
Kaizen, People, Heijunka	

The principle of stopping work immediately, when a problem occurs. (central to Lean)

自 働 化
JI DOU KA
JIDOKA
Autonomation
"automation with a human element"

FIGURE 6.5
Jidoka defined.

It is important to understand the following six key concepts (principles) of jidoka to implement it fully:

- Quality is built in the product or service at the source
- No defect passed forward
- Man/machine separation where operator runs more than one machine at a time

- Multiprocess handling—multitasking
- Self-detection of errors to prevent defects
- Stop and fix

INTEGRATE IMPROVEMENT WITH WORK

In a Lean organization, *all* employees have their work divided into two parts: routine work and improvement work.

Lean culture promotes employee empowerment. With empowerment comes the feeling of ownership and ownership begets the responsibility to improve continuously. Employees need to be developed as process detectives investigating the current state of the process to look for opportunities to improve.

To do this, employees need information (they need to be communicated constantly), tools, and support to implement improvements as they identify them.

Table 6.3 depicts the responsibility to improve among employees at different levels.

SEEK PERFECTION

A Lean organization relentlessly pursues perfection. Therefore, these organizations will remain in the forefront and never become dormant.

TABLE 6.3

How Employees at Different Levels Can Integrate Improvement with Work

Level	Responsibility
Senior leader	Improvement of strategy planning processes and resource allocation
Middle management	Improvement of quality systems, product/service flow, employee development, training, and communication
Shop floor–front line employees	Improvement of cycle times, quality, standard work, visual management, error proofing (poka yoke), etc.

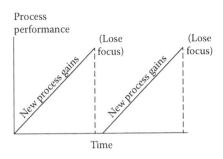 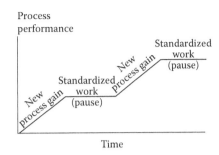

FIGURE 6.6
Standardized work sustains improvements.

Organizations can achieve a state of continuous improvement in two ways:

1. Kaizen (continuous "incremental" improvements)
2. Kaikaku (radical "breakthrough" improvements)

Kaizen combined with breakthrough improvement is the better approach to take (Figure 6.6).

7

Quality Is Built in the Product or Service at the Source

Quality is built in the product or service through the following nine elements:

1. Standard work
2. Successive checks
3. Self-checks
4. Visual management
5. Poka yoke
6. Separate man from machine
7. Multiprocess handling—Multimachine handling
8. Cellular (Cell) manufacturing
9. Stop and Fix

STANDARD WORK

In the tools and techniques section, standard work has been described in detail. It is useful to understand how vastly important standard work is in Lean implementation.

During the entire process of providing the product or service, there are points where tasks critical to customers are performed. It is important for workers to know what these critical points are.

Now the workers are not supposed to remember every critical task. Therefore, it is very important to document what to inspect, how to inspect, and what to do when a defect occurs. The document should also describe how to carry out the task "right at the first time and every time."

SUCCESSIVE CHECKS

These are the inspection checks performed by the operators on the tasks performed by the operators upstream in the value stream (customer-focused process). This method is also known as next operator as a customer (NOC). The operator receiving the operation checks the work of the previous operator or the supplier operator. Successive checks provide immediate quality feedback. These checks also prevent value addition to faulty product or service information. Without successive checks, it is difficult to drive timely root cause analysis and continuous improvement.

SELF-CHECKS

These are performed by the operators following the standard work. Self-checks prevent defective product or service going to the next process step.

Successive checks and self-checks are known as source inspection. Skill training is the key here because the operator should know not only his or her operation but also the upstream and downstream operation. Source inspection keeps defects from occurring during the work.

VISUAL MANAGEMENT

Visual factory is a term to describe how data and information is conveyed in a Lean manufacturing environment. In a Lean manufacturing environment, time and resources dedicated to conveying information, although necessary, are a form of waste. By using visual methods to convey information such as signs, charts, and andons, information is easily accessible to those who need it. The current status of all processes is immediately apparent.

Andon is a Japanese term meaning paper lamp. Andon refers to any visual display that enables operators to signal a line status based on color: green for normal operation, yellow when assistance is needed, and red when the line is down. Nowadays, more sophisticated visual displays are used for andons, but their purpose—efficient, real-time communication of plant floor status—remains the same (Figure 7.1).

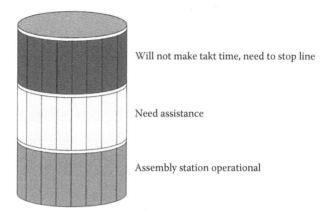

Will not make takt time, need to stop line

Need assistance

Assembly station operational

FIGURE 7.1
Modern andon panel.

We will discuss visual workplace in the Chapter 8, item 1. It is enough to remember that the visual factory principle is a very important aspect of Jidoka.

POKA YOKE (ERROR PROOFING)

In Japanese, *poka* means "inadvertent error" and *yoke* means "prevention."

Error proofing is a systematic approach for anticipating and detecting potential defects and preventing them from reaching the customer (internal or external).

We will have more information on error proofing in the Chapter 9, item 5.

No Defect Passed Forward

This principle is the same as self-checks. This principle is a key to one-piece flow manufacturing because passing the defects along will disrupt the flow. For JIT, this principle is also extremely important.

SEPARATE MAN FROM MACHINE

Figure 7.2 presents the evolution of autonomation.

The evolution towards Jidoka

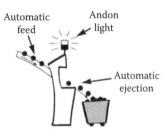

Manual feed and watch machine cycle.

Watch machine cycle.

Self-monitoring machine.

Manual process. One operator for one machine. Process is labor intensive. Use VSM and continuous flow tools for going to next stage.	Mechanization. Operator interaction needed to check process and load/unload operation.	Autonomation. Machine stops when defect is detected or the set quantity is produced. Poka Yoke makes the machine incapable of making a mistake, or accepting a bad part.

FIGURE 7.2
Stages of manual process, mechanization, and autonomation.

MULTIPROCESS HANDLING— MULTIMACHINE HANDLING

Cross-training is a prerequisite for this principle. The work practice of assigning operators to operate more than one process in a product flow-oriented layout requires training operators to operate different types of machines (e.g., blanking, crimping, and inspection).

This practice contrasts with the typical mass production practice of placing operators in separate departments such as turning, milling, grinding, etc., where they work only one type of machine and make batches of parts to transfer to other processes in other departments.

CELLULAR (CELL) MANUFACTURING

The organization that implements multiprocess handling and cellular manufacturing can improve quality by reducing defects and improve

Cellular flow layout

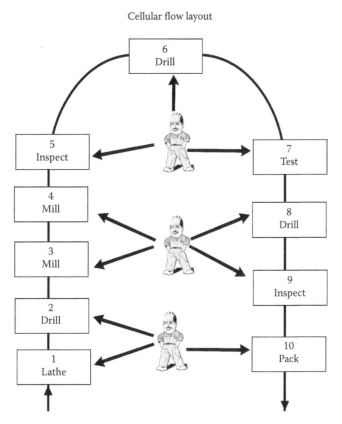

FIGURE 7.3
Typical manufacturing cell layout.

man and machine utilization (see Figure 7.3). Job enrichment promotes employee satisfaction.

STOP AND FIX

This is the extension of the Jidoka principle, where the machine stops until the immediate solution is found. This is followed by root cause analysis and preventive solution implementation to prevent recurrence.

As soon as you read this, a thought comes to mind, "You can't disrupt the process. It is expensive." This Lean principle dictates that as soon as a defect is detected, production is stopped, root cause is identified, and corrective action taken to prevent the reoccurrence before the production is restarted. It may be easily appreciated that half to one hour spent in downtime to "stop and fix" the issue now will save a significant amount of value in the long run.

If a problem is detected that cannot be solved during the work cycle and an operator discovers a problem with parts, tools, materials supply, safety conditions, etc., the operator pulls a rope or pushes an andon button to signal the supervisor. The supervisor assesses the situation and determines if the problem can be fixed before the end of the current work cycle. If the problem can be fixed, the supervisor resets the signal system so the line does not stop. If the problem cannot be corrected within the remainder of the cycle time, the line stops at the end of the work cycle.

Section III

Continuous Improvement
Process Tools and Practices

8

Continuous Improvement Process System

The system can be defined as "a group of interdependent processes and people (also known as subsystems) that together perform a common mission."

ASQ Glossary

Continuous improvement principles can be effectively followed by installing continuous improvement systems.

These systems provide

- Clear communication channels
- Accountability
- Planning for current and future states of the value addition process
- Tracking of planned activities
- Actionable observations and countermeasures when needed

Figure 8.1 is taken from the Malcolm Baldrige National Quality Award (MBNQA) criteria. It shows us how a system approach leads an organization to achieve excellence. The lines with arrowheads represent the interdependent processes, including the people in each process. The system is ineffective as long as it is

- Reacting to the problems OR firefighting
- In the early stages of process development; lacking an integrated approach
- An aligned approach where processes are repeatable—all departments pulling in the same direction to fulfill organizational goals

It's a system...

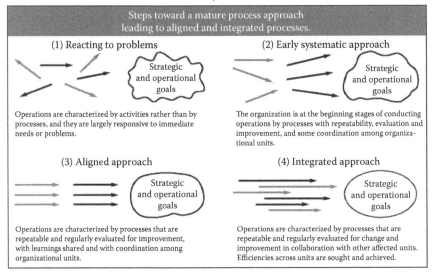

FIGURE 8.1
Systems approach.

- An integrated approach of a multiunit organization where efficiencies across all units are sought, shared, and achieved

For Lean transformation, the continuous improvement process systems are

1. Visual workplace
2. Lot size reduction and load leveling
3. Total productive maintenance (TPM)
4. Standard work
5. Continuous improvement
6. Corrective action system

VISUAL WORKPLACE

Without visual signs, our airports, railway stations, hospitals, and our national highways would be in chaos. As a matter of fact, in any place where human beings gather around to work—whether it is a factory, service center, or an insurance office—the aim is to make work safer, simpler, more efficient, and economical by using visual aids. These aids, signs, or pictures are better than a thousand words.

Gwendolyn D. Galsworth, Ph.D., author of *Work That Makes Sense* (Visual-Lean Enterprise Press, 2010) and *Visual Workplace/Visual Thinking* (Visual-Lean Enterprise Press, 2005), and recognized visual expert, defines a visual workplace as follows:

> A visual workplace is a self-ordering, self-explaining, self-regulating, and self-improving work environment… where what is supposed to happen does happen on time, every time, day or night—because of visual solutions.

This definition describes the results after implementing the visual workplace system. When a workplace gets visual, it becomes safer, better, faster, cheaper, and smoother. All value streams of the products and services flow without disruptions from wastes.

Transferring Vital Information into Visual Devices

A visual device provides vital information to control and influence behaviors. It can be in the form of a tool, picture, or an electronic device that guides or controls behavior by making information vital to the task at hand easily available at a glance; silently, without speaking a single word.

Figures 8.2 through 8.13 provide some visual workplace examples.

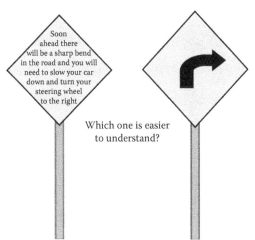

FIGURE 8.2
A picture is better than a 1000 words.

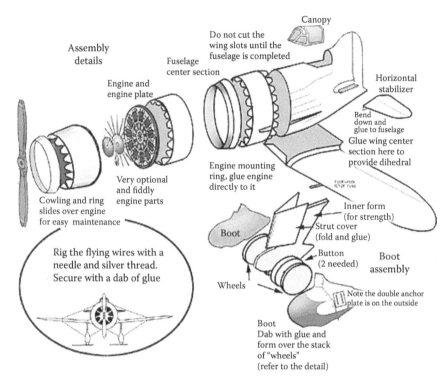

FIGURE 8.3

Model airplane assembly pictorial instruction.

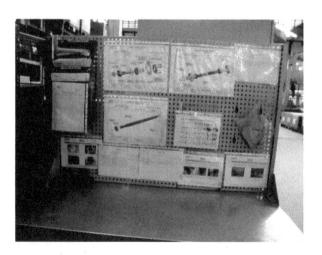

FIGURE 8.4

Visual work instruction.

Example of "order"

Before   After

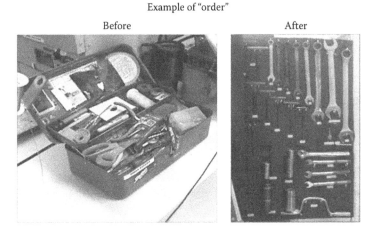

FIGURE 8.5
Use of a shadow board for easy access to the hand tools.

FIGURE 8.6
Visual device showing work activities.

"The First Question Is Free" Rule of the Visual Factory

You may have come across supervisors and managers complaining that their main job is to provide answers day in and day out! Well, here is a visual system answer that will minimize such interruptions and make the workplace doubt- and conflict-free, resulting in an efficient and reliable workplace.

Whether you are a quality executive, marketing manager, technician, accountant, doctor, or a nurse, here are the steps to implement it:

FIGURE 8.7
Visual devices providing ease of access.

Standard work—safety

Safety audits   Aisles way markings

FIGURE 8.8
Safety audits and safe aisles.

When a person comes and asks you a question, answer it very politely and clearly. After that person goes away, either write it down or make a mental note: "that's number one."

Then, when that question is asked again by the same person or someone else, answer it very politely and clearly. After that person goes away, either write it down or make a mental note: "that's number two."

Standard work—safety

Signs and zones

Safety zones, fire extinguisher
signs

Hearing protection
areas

FIGURE 8.9
Visual safety signs.

Where do I get my data for quality operating system?

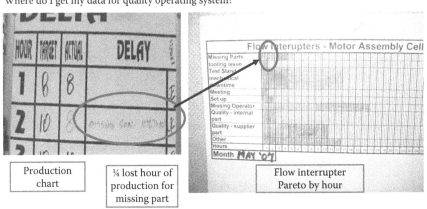

| Production chart | ¼ lost hour of production for missing part | | Flow interrupter Pareto by hour |

Operators should own the quality operating system (QOS) data tracking.

FIGURE 8.10
Visual guide to present data for quality system improvement—1.

Where do I get my data for QOS?

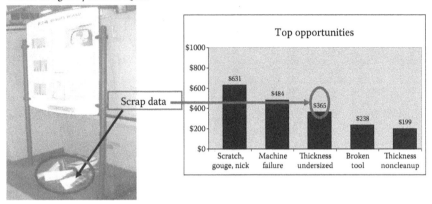

Operators should own the QOS data tracking.

FIGURE 8.11

Visual guide to present data for quality system improvement—2.

Visual accountability board

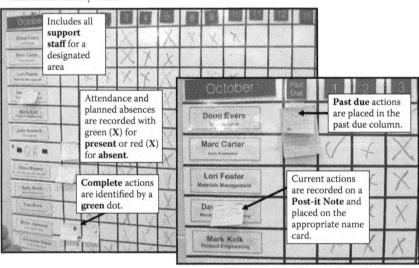

FIGURE 8.12

Visual accountability board.

Standard work—quality

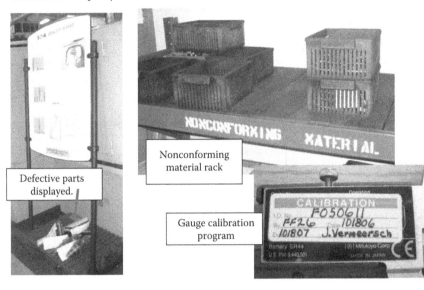

Defective parts displayed.

Nonconforming material rack

Gauge calibration program

FIGURE 8.13
Visual storage, segregation, and identification of nonconforming material.

The first question is free! No action needs to be taken. However, when you hear the same question a second time, it is time to create a visual device so you never ever have to answer that question again and no one has to ask again!

5S STANDARDS SYSTEM COMPLETE WITH IMPLEMENTATION GUIDELINES

After standard work, the 5S system is considered the core process for organizations pursuing Lean. It strengthens the Lean culture and prepares the organization for the Lean journey. Therefore, 5S is the foundation for a disciplined approach to the workplace. 5S can also be considered as five steps.

How did the word 5S come about? 5S originated in Japan; the words in the bracket show the Japanese equivalent.

Step 1: Sort (seiri)
Step 2: Straighten (seiton)
Step 3: Shine (seiso)

Step 4: Standardize (seiketsu)
Step 5: Sustain (shitsuke)

5S Activities

- *Sort*—involves sorting through the contents of the workplace and removing unnecessary items
- *Straighten*—involves putting the necessary items in their place and providing easy access
- *Shine*—involves cleaning everything, keeping it clean daily, and using cleaning to inspect the workplace and equipment for defects
- *Standardize*—involves creating visual controls and guidelines for keeping the workplace organized, orderly, and clean
- *Sustain*—involves training and discipline (5S audit) to ensure that everyone follows the 5S standards

The Benefits of 5S Are

- Safety
- Efficiency
- Quality
- Eliminate waste
- Control over the workplace

Safety

Safety is everyone's concern. When 5S is applied, the safety problems which have thus far been ignored will surface. This happens because:

Regularly
- Guards, covers, and shields are checked
- Electrical fittings and switch gear are looked at closely
- Building and plant conditions are examined

Efficiency

- 5S focuses on the smaller issues so that larger problems are eliminated
- The factory becomes less crowded
- Time isn't wasted in searching for tools and parts
- Production flow is improved

Quality

A 5S Organized Workplace:

- Produces better quality by allowing people to concentrate on performing work without unnecessary walking, reaching, hunting, etc.
- Arranges materials and tools properly to reduce variation in the process and enhances repeatability (standardized work)

Eliminate Waste

When equipment is clean, we can see the waste quickly; this is an improvement in visual control. Addressing waste elimination is a team effort. 5S is considered a fundamental tool to achieving a heightened maintenance system and implementing TPM.

Control Over the Workplace

5S will provide a sense of accomplishment for the employees and empower them by allowing them to

- Identify what is needed in their work area
- Determine where items should be located
- Keep things that way

PDCA MODEL FOR 5S IMPLEMENTATION

The following Shewhart (or Deming) PDCA cycle is a simple, yet powerful, approach for developing, testing, and implementing changes (Figure 8.14). The use of the PDCA cycle brings the power of the scientific method to implement each stage of 5S.

When to Use the Plan–Do–Check–Act

- As a model for continuous improvement
- When starting a new improvement project
- When developing a new or improved design of a process, product, or service
- When defining a repetitive work process

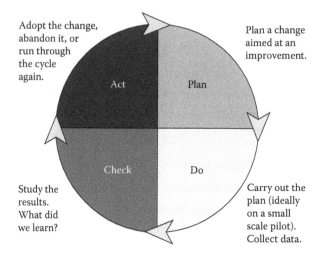

FIGURE 8.14
Plan–do–check–act.

- When planning data collection and analysis to verify and prioritize problems or root causes
- When implementing any change such as the introduction of a 5S system

Plan–Do–Check–Act Procedure

- Plan—recognize an opportunity and plan a change
- Do—test the change. Carry out a small-scale study
- Check—review the test, analyze the results, and identify what you've learned
- Act—take action based on what you learned in the study step

If the change did not work, go through the cycle again with a different plan. If you were successful, incorporate what you learned from the test into wider changes. Use what you learned to plan new improvements, beginning the cycle again (Table 8.1).

5S Audit Radar Chart

- Active participation from the leadership team (Figure 8.15)
- Emphasize starting and sustaining the 5S program

TABLE 8.1

Five Levels of Implementation of 5S

	Sort	Straighten	Shine	Standardize	Sustain
Level A Start	Criteria for disposal of not-needed items have been established	Decide and organize *where* to keep necessary items	Clean work area	Establish work groups to develop and document standardized work methods	Area inspections are random and completed by work groups and/or shop floor leaders
Level B Focus on basics	Tag and identify not-needed items	Decide and organize *how* to keep necessary items	Develop cleaning assignments and checklist	Work methods established for Ss one through three	5S level is established and posted in the area
Level C Make it visual	Establish a safe and secure holding area for not-needed items	Make it easy	Inspection during cleaning	Documented and standardized controls for Ss one through three	Work area 5S is maintained
Level D Focus on repeatability	Evaluate and remove unnecessary items from the site	Make it obvious	Cleaning is an everyday part of the job	Develop standardized work procedures for all work areas	Internal inspection process is standard
Level E Continuous improvement	Identify problem areas and document prevention actions	Identify problem areas and document prevention actions	Identify problem areas and document prevention actions	Share best practices with internal and external work groups	Root cause problem solving process in place

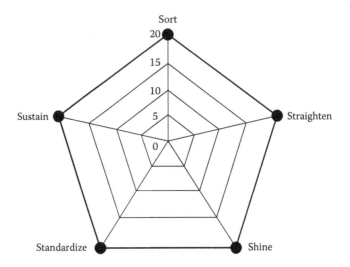

FIGURE 8.15
Radar chart to record audit score of each step of 5S.

- Conduct 5S audit
- Set a recurring 5S audit schedule (daily, weekly, etc.)
- Recognize and reward results

Now let us consider each step in detail.

SORT

Key Information for Sorting

- Sort through everything in the work area
- Separate the items that are not needed or are in the wrong place
- Remove those items from the work area
- "If in doubt, move it out!"

Key Actions to Sort

- Develop criteria for disposal of not-needed items
- Tag and identify not-needed items (include office areas)
- Establish a safe and secure holding area for not-needed items
- Evaluate and remove unnecessary items from the site

Develop Criteria for Sorting

- How often do I need to use the sort operation for not-needed items?
- If it is daily or once a week, keep the hold area near the workplace
- If it is less than once a month, keep the hold area in a remote place
- If it is seldom used, say, once a year or is broken, having no value— consider disposal
- Be reasonable about personal items (food packs, God/family pictures, plants, etc.)

Backpacks, handbags, etc., that can cause safety or quality problems should be kept in lockers or away from the workplace.

"Red tagging" is a visible way to identify items that are not needed or are in the wrong place in the workplace (Figure 8.16).

Questions you should ask yourself (Figures 8.17 through 8.19):

- Is this item needed?
- If it is needed, is it needed in this quantity?
- If it is needed, should it be located here?
- Can you find any unnecessary items cluttering up your workplace?
- Are there tools or materials left on the floor?

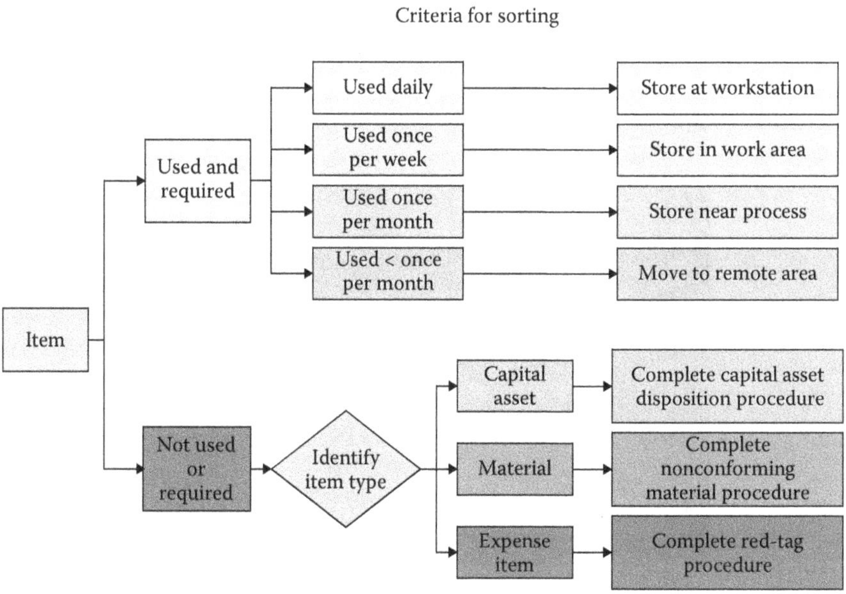

Criteria for sorting

FIGURE 8.16
Red tag procedure.

Red tag		
Date:.. Dept. .... Name		
Condition of item: ☐ New	☐ Used	☐ Broken
Item category		
Expense items		
☐ Tools/fixtures	☐ Production equipment/supplies	☐ Office equipment/supplies
Item description: _____		
Fixed asset	–Follow fixed asset disposition procedure– See accounting department	
To be completed by manufacturing engineering department personnel		
ME input. Evaluated by.. Date:..............................		
Disposition: (manufacturing engineer input)		
☐ Moved to red tag area ☐ Scrap ☐ Fixed asset disposition ☐ ME racks ☐ Repair and use		
Assigned in:	Date:	

FIGURE 8.17

Red tag example.

FIGURE 8.18

Holding area office.

FIGURE 8.19
Holding area equipment.

Move Items to a Holding Area

A holding area is a temporary storage place for red-tagged items that need to be removed from the workplace but cannot be disposed of until all interested parties agree. Holding areas should be "local" for each department. A "central" holding area can be set up for items whose final disposition cannot be decided at the local level. Holding areas should be highly visible and clearly labeled.

Removing Items from the Holding Area

The Holding Area Coordinator is responsible for logging in items as they are delivered, and, finding out the asset's value. The Coordinator will be the contact person who is authorized to categorize the items as follows:

Use the Bottom Gray Portion of the Red Tag and Mark

- Dispose
- Find a new home
- Return to original location

The holding area should be cleared in a timely manner—daily, weekly, monthly (Table 8.2).

When all the criteria for "sort" are satisfied after a thorough audit, give yourself a score according to the score guide chart in Table 8.1, and mark the score on the radar chart (Figure 8.20).

Note: Here, 20 is the perfect score when all check sheet items are completed.

TABLE 8.2

Key Action Checklist for Sort Stage

Key Action	Activity	Responsibility	Check
1. Develop criteria for sorting	Develop criteria for disposal of not needed items		
2. Picture it	Take "before" photos		
3. Tag it	Tag and identify not needed items		
4. Move items to holding area	Establish a safe and secure holding area for not needed items		
5. Remove items from holding area	Evaluate and remove unnecessary items from the site		
6. Picture it	Take "after" photos		
7. Evaluations are conducted	Complete evaluation using "5S" levels of implementation		
8. Stage inspection	Contact plant manager/key management person for evaluation		

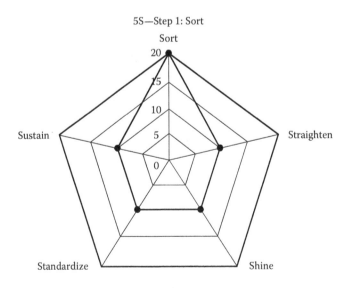

FIGURE 8.20
Sort audit score on radar chart.

STRAIGHTEN

Key Information about Straighten

- Decide where to keep necessary items
- Organize items by frequency of use and clearly designate their correct location
- Make it easy to find and use them
- Make it visually obvious when they are not in their correct place
- "A place for everything, and everything in its place!"

Key Actions to Straighten

- Determine location for every item
- Outline locations of equipment, supplies, common areas, and safety zones with lines (tape)
- Develop shadow boards
- Label needed items
- Determine where and how much inventory or supplies should be kept on hand. Use visual aids like level marks, two bins, and color coding (Figure 8.21)
- Determine location for items
- Use "5 Whys" to decide where each item belongs
- Use the "5 Whys" technique to determine where each item should be located. The idea is to ask "Why?" until you can no longer ask, "Why?" The team then decides where each item belongs
- For an interactive example demonstrating the use of the "5 Whys" (see 5 Whys example in shine)
- Locate needed items so they can be retrieved in 30 seconds with minimum steps

Outline Locations with Lines (Tape)

- Divider lines define aisles/ways and work stations
- Marker lines show position of equipment
- Range lines indicate range of operation of doors or equipment
- Limit lines show height limits related to items stored in the workplace
- Tiger marks draw attention to safety hazards
- Arrows show direction

FIGURE 8.21
5S implementation by operators. (From 5S for Operators, Productivity Press.)

The picture below shows a "tiger" tape and some of the industrial marking tapes. A scraper may be needed if the tape needs to be replaced or shifted. Easy to see if one is missing. Everything is available for setup (Figures 8.22 through 8.26).

When all the above criteria for "straighten" are satisfied after a thorough audit (Table 8.3), give yourself a score according to the radar chart (Figure 8.27).

SHINE

Key Information about Shine

- Clean everything, inside and out
- Find ways to prevent dirt, oil/water leaks, and contamination from occurring
- Adopt cleaning as a form of inspection
- Make cleaning part of everyday work

FIGURE 8.22
Tape used to outline various locations.

FIGURE 8.23
Example showing the outlines of locations.

Note: Shine and step 1 of TPM (autonomous maintenance) are very similar in that they both focus on cleaning and inspecting the equipment.

The most obvious goal in the cleaning step is to turn the workplace into a clean, bright place where everyone will enjoy working. Another key purpose of step 3 is to keep everything in top condition so when someone needs to use something, it is available and ready to use. By implementing step 3, "spring-cleaning" or "year-end cleaning," or "Diwali cleaning in India" becomes a thing of the past as cleaning becomes a part of daily work habits.

FIGURE 8.24
Develop shadow boards.

FIGURE 8.25
Items on the right needed labeling and are properly labeled on the left.

Immediate Benefits of Cleanliness

- Well-lit, clean environment; improved quality
- Absence of puddles of oil and water improves safety
- Machines will not break down as frequently when they are cleaned, maintained, and inspected
- Improved delivery to our customers by reducing lead time
- Machines will operate consistently and correctly with regular maintenance

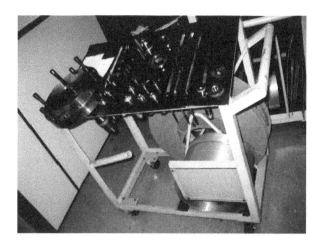

FIGURE 8.26
Develop tool carts.

TABLE 8.3

Key Action Checklist for Straighten Stage

Key Action	Activity	Responsibility	Check
1. Picture it	Take "before" photos		
2. Determine location for items	Decide and organize where to keep necessary items		
3. Outline locations with lines (tapes)	Outline locations of equipment, supplies, common areas, and safety zones with lines (tapes)		
4. Develop shadow boards	Install shadow boards. "A place for everything and everything in its place"		
5. Label needed items	Identify all needed items with labels		
6. Picture it	Take "after" photos		
7. Evaluations are conducted	Complete evaluation using "5S" levels of implementation		
8. Stage inspection	Contact plant manager/key management person for evaluation		

- Reduced downtime and improved quality
- Cuttings and shavings will be cleaned away, and not mixed into production and assembly processes; improved quality
- Fewer injuries will result from cuttings and shavings being blown into people's eyes; improved safety

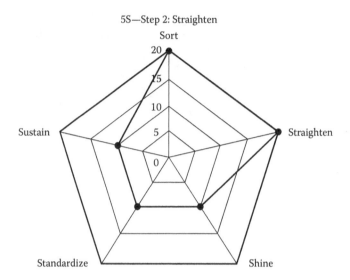

FIGURE 8.27
Straighten audit score on radar chart.

Key Actions to Shine

- Perform initial cleaning of workplace and equipment (Figure 8.28)
- Find ways to keep everything clean
- Adopt cleaning as an opportunity to improve productivity through value-added form of inspection
- Make cleaning part of everyday work for all employees
- Perform initial cleaning of equipment (Figure 8.28)
- Identify and tag any items that you feel may cause oil/water leaks and contamination

Before   After

FIGURE 8.28
Find ways to keep everything clean.

- Analyze reasons for contamination and brainstorm countermeasures or solutions to these problems
- Use the "five whys" and "cause and effect diagram" techniques to brainstorm the root cause of a problem
- To analyze reasons for contamination, use brainstorming techniques like "five whys" and "fishbone analysis" to help find the root cause
- Develop strategies to eliminate sources of contamination, like controlling coolant overspray, and containing cutting shavings as they are produced
- Ask the five whys (Figure 8.29)
- Cause and effect diagram (Ishikawa diagram or fishbone diagram; Figure 8.30)
- Find ways to keep everything clean (Figure 8.31)
- Keep a log of areas to be improved
- Example of a simple log to be kept by the area team leader/supervisor (Table 8.4)

The One-Minute or One-Point Lesson

This display conveys immediate knowledge about a single process, problem, or skill that should be shared among the teams and also among various shifts. The display contained here has a brief instruction on what a one-point lesson (OPL) is, its objective, and how it is generated. In Lean organizations, an open binder for archived lessons on OPL is maintained for all to refer to and get more ideas for continuous improvement.

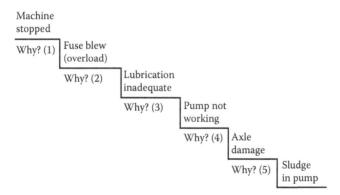

Corrective action: install a filter on the lubrication pump.

FIGURE 8.29
Five whys exchange.

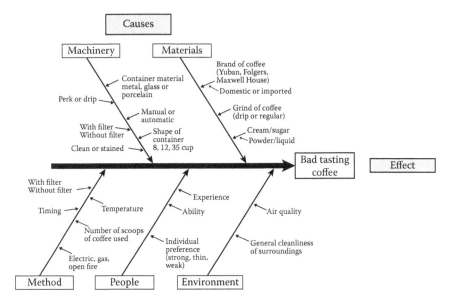

FIGURE 8.30
Cause and effect diagram.

Lean one minute lesson
Title: Table saw vacuum—SWBD pallets

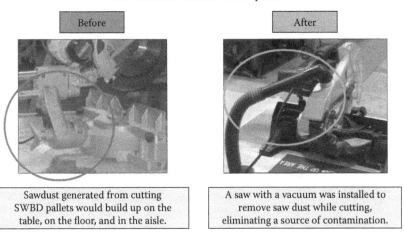

Before	After
Sawdust generated from cutting SWBD pallets would build up on the table, on the floor, and in the aisle.	A saw with a vacuum was installed to remove saw dust while cutting, eliminating a source of contamination.

FIGURE 8.31
Table saw with and without vacuum saw dust extraction.

TABLE 8.4

Follow-Up Action Items Log

Location of Contamination	Problem	Solution	Who	When
On the floor around hydraulic press No. 3	Oil overflowing from drip pan	Replace gasket	Anil	5/15/2008

Adopt Cleaning as a Form of Inspection

Cleaning and inspecting keeps dirt and grime off equipment and exposes slight defects. As we clean, we touch the equipment, check it, and discover impending failures. These items are then tagged for maintenance attention.

- This step is a precursor to developing a formal TPM program
- Make cleaning part of everyday work for all employees
- Cleaning builds pride in the workplace
- Operators are encouraged to use their five senses to catch problems while they are small

Develop daily check sheets as shown in Figure 8.32.

Important Note for the Team Leader

Have the team complete "shine" activities by using the "key action check sheet" (Table 8.5). The "5S levels of implementation" matrix is found in the Introduction. The plant manager (or staff member) should use this matrix to evaluate compliance with all five levels of shine. Upon successful compliance, the plant manager will attach a gold seal to shine on the 5S wheel that is part of the activity board. See example in Introduction.

Time is very important in conducting "shine." Monitor the time closely, and keep the team on track. When satisfied, complete the radar chart (Figure 8.33).

5S checksheet

Area:____________________ Supervisor:____________________ Month: ____________

Photo #		M	T	W	H	F	S	M	T	W	H	F	S	M	T	W	H	F	S	M	T	W	H	F	S
	Daily activities																								
1		A	B	C	D	E	A	B	C	D	E	A	B	C	D	E	A	B	C	D	E	A	B	C	D
2		B	C	D	E	A	B	C	D	E	A	B	C	D	E	A	B	C	D	E	A	B	C	D	E
3		C	D	E	A	B	C	D	E	A	B	C	D	E	A	B	C	D	E	A	B	C	D	E	A
4		D	E	A	B	C	D	E	A	B	C	D	E	A	B	C	D	E	A	B	C	D	E	A	B
5		E	A	B	C	D	E	A	B	C	D	E	A	B	C	D	E	A	B	C	D	E	A	B	C
	Weekly activities																								
6						A						B						C						D	
7						B						C						B						E	
8						C						D						E						A	
9						D						E						A						B	
10						E						A						B						C	
	Monthly activities																								
11																								ABCDE	
12																								ABCDE	
13																								ABCDE	
14																								ABCDE	
15																								ABCDE	
																								ABCDE	
	Supervisor signoff (weekly)																								

Assignments

A	
B	
C	
D	
E	

FIGURE 8.32
5S check sheet for daily, weekly, and monthly activities.

TABLE 8.5

Key Action Checklist for the Shine Stage

Key Action	Activity	Responsibility	Check
1. Picture it	Take "before" photos		
2. Perform initial cleaning of workplace	Clean everything inside and out		
3. Find ways to keep workplace clean	Develop team problem solving skills. Analyze contamination sources		
4. Adopt cleaning as a form of inspection	Check for contamination while cleaning, and document findings		
5. Make cleaning a part of everyday work	Develop activity check sheets and divide responsibilities		
6. Picture it	Take "after" photos		
7. Evaluations are conducted	Complete evaluation using "5S" levels of implementation		
8. Stage inspection	Contact plant manager/key management person for evaluation		

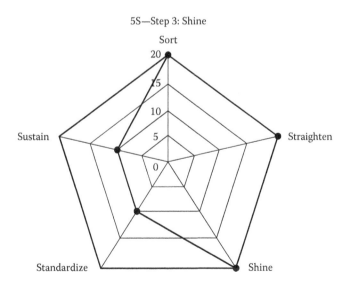

FIGURE 8.33
Shine audit score on radar chart.

STANDARDIZE

Key Information about Standardize

Many organizations pursuing 5S often say that they failed at sustaining their efforts—the fifth "s." However, it is more likely that they failed at "standardize." Establish the routines and standard practices for regularly and systematically repeating the first three Ss.

Create procedures and forms for regularly evaluating (grading) the status of the first three Ss. An important element here: Don't create a rule that no one will follow.

Key Actions to Standardize

- Document how organizing and cleaning will be done and make standards visible by means of photos, videos, etc.
- Maintain and monitor the first three "Ss"
- Standardize red-tag procedures and holding area rules (see Sort)
- Standardize procedures for creating shadow boards, position lines, and labeling of all items (see Straighten)
- Standardize cleaning schedules using the "5S operator check sheets" (see Shine)

FIGURE 8.34
Standardize tool inspection area.

FIGURE 8.35
Standardize office area.

- Standardize "one-minute lessons" for documenting and communicating 5S procedures and improvements in workplace and equipment
- Maintain and monitor the first three Ss

The methodology of by whom and how often to inspect should reflect the culture of the organization. A common approach is for cross-functional teams to inspect areas on a weekly cycle (Figures 8.34 and 8.35; Table 8.6).

When satisfied, complete the radar chart (see Figure 8.36).

SUSTAIN

- Record and report 5S activity for lessons learned and knowledge management
- Take "before" and "after" photos
- Document the results using a standard form to create a summary report folder
- Submit the summary report to the Lean coordinator
- Review summary report for possible best practices
- Upload the summary report to the best practices website
- Use knowledge management to share and begin sustaining 5S in another area of the same plant or another office/warehouse/plant (Figures 8.37 and 8.38)

TABLE 8.6

Key Action Checklist for the Standardize Stage

Key Action	Activity	Responsibility	Check
1. Picture it	Take "before" photos		
2. Maintain and monitor the first three Ss	Sort, straighten, and shine are fully implemented and functioning well		
3. Document how organizing and cleaning will be done and make standards visible	Post all team activity documents/checklists on the team 5S board for visibility		
4. Perform internal audits	Audit the team to validate success		
5. Picture it	Take "after" photos		
6. Evaluations are conducted	Complete evaluation using "5S" levels of implementation		
7. Stage inspection	Contact plant manager/key management person for evaluation		

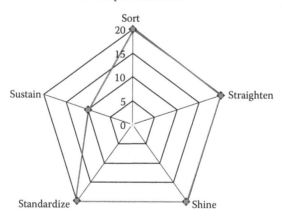

FIGURE 8.36
Standardize audit score on radar chart.

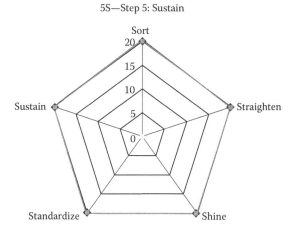

FIGURE 8.37
Sustain audit score on radar chart.

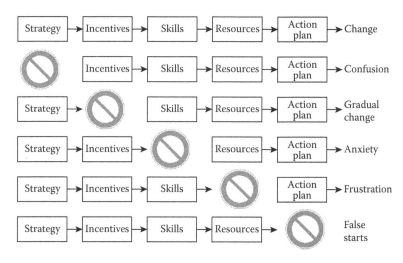

FIGURE 8.38
Requirements for a change.

Lesson

- Strategy with incentive/reward, skill development, providing required resources, making an action plan that will bring about change
- With any one of these five aspects missing (shown by a "no entry" sign), the result will be either confusion, lethargy, anxiety, frustration, or false starts!
- Use 5S to begin Lean evolution and to sustain forward progress

- 5S is a fundamental part of TPM
- 5S is a fundamental part of standardized work

Five Critical Factors Ensure a Successful 5S Implementation

- A customer-centric culture
- Positive financial results obtained during initial 5S implementation
- Management engagement and encouragement
- Strong resource commitment by the top management
- An infrastructure that integrates Lean Six Sigma projects into the organization's real work

The 5S system audit process is shown in Figure 8.39.

LOT SIZE REDUCTION AND PRODUCTION LEVELING SYSTEM (HEIJUNKA)

Heijunka Definition

The distribution of production volume and mix evenly over time is defined as *heijunka*. The heijunka system is based on two important concepts

- Takt time
- Changeover time reduction or setup reduction

In traditional push production operation, big lots or batches of product are created and "pushed" to the next process, regardless of whether the product is needed. At this point, the product waits in a queue as work in process (WIP).

Normally, large batches are created as a result of "forecasted" customer demand. This results in finished goods inventory. The idea is to have the product on hand to meet the forecasted demand if and when the customers place orders.

To eliminate this type of fluctuation (mura), load leveling, also called production leveling, has been successfully applied in many small and large production organizations.

5S Audit		*Insert location address here*	# of obs	points	Rev #	New
			0	4	Rev date	20/5/08
			1	3		
			2	2	**Points**	
Prepared by		Date	3 a 4	1		
Department	Production		>4	0		

Maximum points per category = 20 points
Put the number of points in the square

Category	Item	S/A U/A S/M Tst	Comments	Area audited
Sort (Trash the unneeded) (Daily vs. monthly)	**Differentiate between what is necessary and what is not necessary.**			
	All items which are unnecessary for production should be removed from the work area. All personal items should be placed inside a locker or a cabinet.			
	Tools for daily use should be within the easy reach of the operator.			1 Subassembly
	The tools used weekly or monthly should be placed away from the work place to avoid clutter.			
	Attach a red tag to identify the items which are not necessary in each area and remove them at least once a month.			2 Unit assembly
	Interview 4 persons in the area—what are the steps you will take if you find an unnecessary item or object in your area?			3 Super market (Kitting)
Straighten (Organize)	**A place for every thing and every thing in its place.**			
	All locations of materials/bins are clearly identified using colored tapes according to the color code.			4 Test area
	All trolleys and cabinet drawers for tools to have foam shadeboards and no toll should remain out of its place.			
	Nothing should be above the assembly units or loose in the drawers.			
	All documents and records are clearly marked and stored and protected.			
	Locations of all incoming materials in the work areas are marked and located in such a way that they do not obstruct the passals and isles.			
Shine (Clean1)	**Clean and keep every thing clean.**			
	The visual controls like labels and tapes used in marking are clean and in good condition.			
	Paint and/or keep all machines, presses, test stands, accessories and utilities (switches, air outlets, pressure controllers, etc.) clean and ensure that there are no air or oil leaks.			Area supervisor
	All work areas, meeting areas, passages and floors are clean.			
	Cleaning materials and implements are easily accessible.			
	Interview 4 persons of the area if they know what is their time table for cleaning?			
Standardize (Train, document and record)	**Clean and look out for methods of keeping everything clean.**			
	All employees should be trained in 5S.			Auditor
	A notice board in the are should display the improvements made each day or each week.			
	All area improvements should be summarized in the department notice board each week.			
	The notice boards will contain Radar Charts, and trend charts of the 5S categories with responsible persons and completion dates.			
	All employees should wear clean uniforms and have their IDs.			
Sustain (Discipline and training)	**Constantly comply to the 5S rules.**			
	Number of criterias missing in the category Sort			
	Number of criterias missing in the category Straighten			
	Number of criterias missing in the category Shine			
	Number of criterias missing in the category Standardize			
	The actions for nonconformities found in previous audit are completed.			
	Total points of the are audited.			

3 opportunities of improvment in the area	Completion date	Responsible	
1			5S audit form for production area.
2			
3			

Revision #1 20-03-2014

FIGURE 8.39
5S audit form.

There are three components to production leveling (Figure 8.40):

1. Total volume of all models
2. Model sequence in which the models are required
3. Volume of each model

Total Volume of All Models

Customer demand varies. However, there are predictable patterns over specific periods over which the demand, on average, is relatively constant (Figure 8.41; Table 8.7).

Model Sequence and Model Volume

As the day-to-day customer demand varies, so does the type of product or service. The realization of these products and services demands time

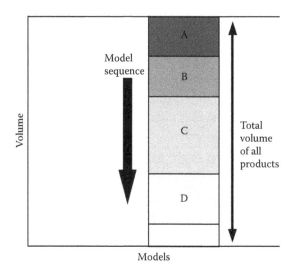

FIGURE 8.40
Components of production leveling.

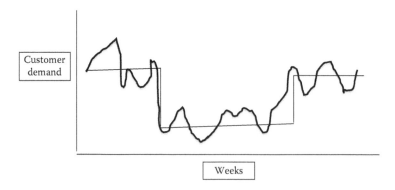

FIGURE 8.41
An example of production leveling—1.

needed for changeovers from one type to the other. This affects the overall capacity of the organization. A reduction of the changeover time is the key factor here. *Changeover time begins when the last good part or service is completed and ends when the first good part or service is started at normal cycle time and efficiency.*

Therefore, most Lean companies strive hard to reduce the changeover time. Generally, the common scenario in product/service provision is that of mixed model production.

TABLE 8.7

An Example of Production Leveling

Week	Total Demand	4-Week Average	Weekly Level Production
1	478	434	434
2	376		434
3	453		434
4	432		434
5	362	423	423
6	480		423
7	450		423
8	400		423
9	350	407	407
10	440		407
11	450		407
12	390		407

Equal Volume but Mixed Products

A customer needs 20 units per hour of product A, 20 units per hour of product B, and 20 units per hour of product C. This means a total number of 60 units need to be produced per hour for products A, B, and C.

The takt time for product is:

$$\text{Available time/required volume} = 60 \text{ minutes}/20 + 20 + 20 =$$
$$1 \text{ product per minute.}$$

This means the product sequence can be:

ABCABCABC, where products A, B, and C are produced at intervals of three minutes each. With a sequence of AAABBBCCC, the customer for product C will have to wait a full nine minutes to receive product C, and six minutes to receive product B. The customer for product C can't wait that long and goes to another supplier, say Sub-Way, leaving McDonald's causing it a loss of business and possible increase in WIP inventory buildup.

Different Volumes and Mixed Products

A customer needs ten units per hour of product A, 20 units per hour of product B, and 30 units per hour of product C. This means a total number of 60 units need to be produced per hour for products A, B, and C.

Product A: Demand/total demand = 10/60 = 1/6
Product B: Demand/total demand = 20/60 = 1/3 = 2/6
Product C: Demand/total demand = 30/60 = 1/2 = 3/6.

Determining the least common denominator, we will be able to find the repeatability number. The repeatability number (denominator) is the total number of open slots in the sequence, and the numerator is the number of slots in the sequence.

Product A has denominator 6, or every sixth slot. Its numerator is 1 so it has one slot. Product B has 2 slots, or every third slot. Slot number 6 is taken so B will occupy slot number 5. The third slot from 5 is slot number 2. Therefore, product 5 occupies two slots (5 and 2). In the end, product C occupies the rest of the slots.

Product	C	B	C	C	B	A
Sequence slot	1	2	3	4	5	6

TOTAL PRODUCTIVE MAINTENANCE SYSTEM (TPM) AND ITS IMPLEMENTATION

What is it?

- It is a method to make capital equipment more productive by eliminating and controlling downtime, machine slowdowns, and increasing process capabilities
- TPM is a Lean tool to optimize the effectiveness of manufacturing equipment and tooling
- Starts with 5S/visual factory system implementation
- Builds a comprehensive downtime database by cause, frequency, and duration
- Predicts and prevents downtime by PM (planned maintenance) system
- Expands the role of the operator as the first point of early warning and prevention
- Develops professional maintenance skills

Effective TPM Eliminates the Following Losses

- Equipment breakdowns
- Quality time losses—defects, scrap, and rework
- Losses due to safety issues
- Losses due to mini stoppages
- Slow down (speed) losses due to reduced speed
- Idle time losses, setup time losses
- Maintenance time losses (breakdowns)

What Are the Steps?

- Perform initial cleaning and inspection
- Eliminate the causes of contamination and inaccessible areas
- Develop cleaning, lubricating, and inspection standards
- Provide basic training in the machine's functions and controls to enable effective general inspection
- Enhance and implement autonomous inspection standards and schedules
- Implement workplace organization and housekeeping, employing 5S and visual controls
- Follow-up on all the above steps and advanced improvement activities

What Are the Resource Requirements for Implementation?

- Cross-functional team of operators, maintenance, supervision, and support areas for three to five days
- Machine availability for five to ten days depending on the machine's condition and time of repairs

First Things First

- Prepare the inventory of equipment and tooling so that all equipment is accounted for
- Prioritize equipment for TPM using a cause and effect matrix (Figure 8.42)

Complete job safety analysis (JSA) for each piece of equipment. Deal with critical equipment first and then the rest of the equipment and tooling (Figures 8.43 and 8.44)

Instructions for Completing the Job Safety Analysis

Job safety analysis (JSA) is an important prevention tool that works by finding hazards and eliminating or minimizing them *before* the job is performed and *before* they become accidents. Use JSA for job clarification and hazard awareness, as a guide in new worker training, for periodic contacts and for retraining of senior workers, as a refresher on jobs which run infrequently, as an accident investigation tool, and for informing workers of specific job hazards and protective measures.

Set priorities for doing JSAs: jobs that have a history of many accidents, jobs that have produced disabling injuries, jobs with high potential for disabling injury or death, and new jobs with no accident history.

Select a job to be analyzed. Before filling out this form, consider the following: the purpose of the job—What has to be done? Who has to do it? The activities involved—How is it done? When is it done? Where is it done?

In summary, to complete this form, you should consider the purpose of the job, the activities it involves, and the hazards it presents. If you are not familiar with a particular job or operation, interview a worker who is. In addition, observing a worker performing the job, or walking through the operation step by step may give additional insight into potential hazards. You may also wish to videotape the job and analyze it. Here's how to do each of the three parts of a job safety analysis.

For each piece of equipment and tooling, develop TPM specifications.

Manufacturing Equipment TPM Specification

1. Develop a cleaning and inspection checklist in a format that can be displayed and used by the operator (in addition to that contained in the machine manuals).

2. Develop a lubrication checklist in a format that can be displayed and used by the operator (in addition to that contained in the machine manuals).

3. Mark the proper operating ranges on temperature, pressure, flow, and speed gauges. Proper ranges can be indicated by lines or shaded regions. The result must give the operator a clear understanding of the operating condition of the machine. The preferred method is to use a gauge with green zones for proper range and a red zone for out-of-tolerance ranges.

Cause and effect matrix
Critical equipment

Item #	Equipment	Back up routing available (10/1)	Back up routing available (outsource) (8/2)	Overall volume through equipment (10/3)	Current failures (10/4)	Total		Criteria
1	MDM tester #1923	9	9	9	9	342		
2	Servo tester #1909	3	9	9	9	282		Score Demand hrs
3	Process water chiller	9	9	9	1	262		9 >20
4	160/360 and 1"2"4" tester #1884	3	9	9	3	222		3 >10-20
5	Reverse osmosis system	9	3	9	3	214		1 <10
6	Gear motor tester #819	9	9	1	3	202	Tier 1 Critical equipment	
7	PVH tester #8175	1	9	9	3	202		Score Backup
8	PVH tester #8112	1	9	9	3	202		9 No available
9	PVB tester #1999	1	9	3	9	202		3 1 available
10	420 tester #2011	9	9	3	1	202		1 More than 1
11	Heui test stand #8162	9	9	1	1	182		
12	B2B tester #1838	3	9	3	3	162		
13	Air compressor	3	3	9	1	154		
14	Gear pump tester #1765	1	9	3	3	142		Score Fail per week
16	Gear pump tester #1474	1	9	3	3	142		9 >3 incidents
17	PVE 19/21 tester #1925	1	9	1	3	142	Tier 2 Critical equipment	3 1 to 3
18	Triple test stand #2031	3	9	1	1	122		1 Less than 1
19	PVB tester #8165	1	9	3	1	122		
20	Gear pump tester #820	1	9	3	1	122		
21	Lazer test stand #1	3	9	1	1	122		
22	Lazer test stand #2	3	9	1	1	122		
23	PVH 3 spindle tester #8108	1	9	3	1	122		
24	PVE 12 tester #2014	1	9	1	1	102		
25						0		
Total		340	384	460	200			

FIGURE 8.42
Cause and effect matrix to prioritize equipment for TPM implementation.

JOB SAFETY ANALYSIS

Hazard type (HT)		Job task:	Plant/location:	Equipment #:
			Department:	Analysis done by:
		PPE required:		Reviewed by:
				Date initiated:
				Revised date:
		Tools used:		Chemicals used:

Hazard type (HT)

1. Awkward postures
2. Dermal exposure
3. Electrical hazards
4. Elevated work (>4 ft.)
5. Chemical reaction
6. Fire hazard
7. Flying debris (chips, dust, etc.) falling objects
8. Heavy lifting (manual or mechanical)
9. Inhalation exposure
10. Laceration/ puncture hazard pinchpoint
11. Radiation
12. Slip/trip
13. Thermal (heat or cold)

Sequence of job steps	Potential hazards	CTS*	Safe work procedure	PPE
1.				
2.				
3.				
4.				
5.				
6.				
7.				

Operator	Supervisor	Manager
Operator signature	Supervisor signature	Manager signature

CTS* = Critical to safety (meaning: if safe work procedure is not followed, operator will be significantly hurt or worse)

PPE = Personal protection equipment

FIGURE 8.43

Job safety analysis form.

JSA example
by: Maria Ryan

DEPT:

TITLE: Bag Manufacturing—Welded Line

JOB INSTRUCTION NO.:

Eaton		Elizabeth, NJ Facility
Job task: Welded line	Dept.: Bag manufacturing	
Analysis by (Name/Title): Akshay Hattangdi	Resource contacts: Maria Ryan, Supervisor	Implementation date:
		Revision dates:
Chemicals:	Required PPE: hair net, safety glasses, smock	Required tools: Allen wrench, screwdriver, scissors
SEQUENCE OF BASIC JOB STEPS (In Order)	POTENTIAL HAZARDS	REQUIRED SAFE PRACTICE
Load material	Back strain	Bend knees when lifting
Weld seam	Burns	Keep hands away from welding head
Cut sleeve	None identified	
Weld and cut bottom	None identified	
Ring weld	Burns, cuts	Keep hands away from welding head
Pack	Back strain	Bend knees when lifting; keep pallet close to body

Safety committee member: _____ Leader: _____

Signature/Date: _____ Signature/Date: _____

FIGURE 8.44
JSA example.

4. Ensure all gauges are visible and easily read.

5. Label all equipment components identifying function and purpose.

6. Install permanent line ID tags on all pneumatic, hydraulic, oil, and coolant lines.

7. Mark directions of flow, feed, or rotation on respective devices (motors, conveyor drives, pumps, etc.).

8. Install clear Lexan or Perspex covers on all belt, chain, and coupling drive covers. The operator should be able to visually check the condition of these items.

9. Use color-coded grease caps to protect and designate lubrication types and frequencies. Ensure that all grease and oiling points are identified and easily accessible.

10. Where centralized grease and oiling systems are used, provide a feedback system to ensure that fluid is being dispensed properly. If this feedback system is not installed, provide a PFMEA or other similar document to define your method of ensuring that the system is operating correctly (e.g., How do you know that the system is applying oil/grease as required?).

11. Ensure that there is a metal lubrication schematic plate attached in a prominent position on the machine referencing the identified grease and oiling points (item 10).

12. Ensure that all internal intake filters for hydraulic systems are plumbed to the exterior of the tank and are plumbed with shut off valves to allow the operator to quickly change the filters. All external filters should have visual or electrical contamination notification gauges on them.

13. Ensure that all liquid level gauges have low and high limit marks on them. Proper ranges can be indicated by lines or shaded regions. The result must give the operator a clear understanding of the fluid level.

14. Install low liquid level alarms on all coolant supply tanks.

15. Apply replacement information labels for belts, chains, filters, etc., on respective covers and/or components.

16. Apply nameplates to all electrical cabinets and panels.

17. Ensure that all documentation for machines are translated in both English and the local language (machine manuals, drawings, schematics, etc.).

18. Eliminate all special tools required to operate and maintain the equipment. If there is a tool required for daily operation of the machine, it must be approved by the process engineer prior to runoff.

19. No tools are allowed for fixture adjustments/changes. These deviations must be approved.
20. Ensure that all wiring and cables are as short as possible. There are to be no wires, cables, conduits, pipes, etc. run on the floor.
21. Ensure that all fresh air intake vents for electrical control cabinets and panels have filter frames and filters installed on them.
22. Install filter frames and filters on all electrical motor fan covers. For inaccessible motors, use a flexible hose to bring the filter out to a convenient place for changing by the operator.
23. Ensure that all hand-operated control valves have open/close indications on them. In the "on" position, the handle must be in line with the direction of flow. When in the "closed" position, the handle must be perpendicular (or opposed in some manner) to the flow.
24. For regularly required operation, eliminate any inaccessible and hard-to-reach areas for the operator.
25. Ensure that there is adequate coolant flushing lines within the machine to disperse chips and facilitate daily cleaning of the immediate machining area. The flushing lines must be contained within the machine enclosure (no fittings or lines outside of the sheet metal enclosure). Storage positions within the enclosure must be provided for the flushing line when not in use to keep them clear of the machine operation.
26. No drip pans are allowed under machines or support equipment (hydraulic units, etc.). No recessed areas on the machine or equipment intended for retaining lost fluid are allowed. The sources of leaks must be eliminated. Fluid leaks will be specifically addressed at machine runoffs.
27. Open cavities or areas on the machine enclosure that cannot be wiped clean should be covered with Lexan or sheet metal.
28. Machine controls must be mounted to the machine, not floor standing.
29. If the equipment has subsidiary systems that are not contained within the machine enclosure (chillers, filtration systems, etc.), the supply/control lines must be quickly disconnected (with fluid retained) and the unit must be on wheels to allow for cleaning around the machine. Deviations required, based on size of the subsidiary system, must be approved.
30. Hoses and wires that are not contained within the machine enclosure must be contained in sheet metal tray with access lids or doors. Deviations for subsidiary systems may be allowed to facilitate movement for cleaning.

31. OSHA lockout diagrams should be placed on all machines, and any special accessories (conveyor systems, etc.) showing where energy sources are located, how the machine is to be locked out, operator position relative to the machine, etc., is to be shown.

32. Hydraulic lines going to two-way cylinders are not required to be marked with direction of flow; single-direction flow lines are required to indicate flow direction.

33. Machines subject to vibrations are to be equipped with vibration absorption pads so as not to transmit vibrations to the floor/other machines.

34. Areas of equipment that are subject to high heat (over 100°F (37°C)) and in an area within human reach are to be covered with some type of permanent insulation to avoid injuries. If this is not possible, then appropriate guarding must be installed to prevent injury. This guarding must be designed for quick removal to allow access to the heat area for maintenance. Example: heat treatment oil filters, normally run over 200°F or 93°C but must be changed periodically.

35. Drain lines (from areas such as air regulators) must be provided with a small collection container that is easily accessed by the operator and of sufficient size to contain a shifts' worth of condensation. They are not to be discharged to the shop floor.

36. All energy sources (water, air, hydraulic) are to have an OSHA lockable device installed at the point of machine connection. Similar to a disconnect built in the electrical cabinet.

37. All digital devices are to have a label nearby stating normal operating ranges. For example, a digital gauge to show the temperature of a washer tank—it says 140°F—what is it really supposed to be?

38. Excessive lengths of cords/hoses are to be contained in a roll-up type device mounted up and out of the way for normal use, so as not to create a work hazard. Example, Teach Pendant cord for a robot or gantry.

39. Communication device connections are to be RJ-45 type to allow all machines to be connected to a plant-wide financial information system and placed within two feet of the CNC control monitor.

40. Heavy machine guards/shields are to be provided with lifting eyes or tapped holes for lifting eyes to facilitate easy removal for repairs.
41. All machines with oil reservoirs to be equipped with either sight glasses and/or level indicators of some type.
42. Heavy belt/chain guards are to be hinged to facilitate quick change of belts and chains.
43. Identification plaques are to be permanently attached to the machine, no tape of any type allowed.
44. Excess cable lengths will be allowed up to 12 in. maximum in electrical cabinets.
45. All machine safety/lube placards to be mounted on the back of machines.
46. The machine supplier is to provide an FMEA for their machine and related support equipment.

Implementing TPM

The first step in implementing TPM is to run a pilot TPM on one of the most critical pieces of equipment (Figures 8.45 and 8.46).

- Complete visual factory 5S stages and JSA for this equipment
- Examine each item of the TPM specification

| Operating ranges of gauges labeled | Well-labeled valves and hose and wire |

FIGURE 8.45
Preparation of new equipment for TPM—1.

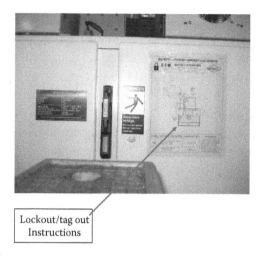

Lockout/tag out
Instructions

FIGURE 8.46
Preparation of new equipment for TPM—2.

- Identify lubricating points. Make changes and improvements as specified in the TPM specification
- Design a form to record lost times

Overall Equipment Effectiveness (OEE)

OEE = availability × performance efficiency × quality rate
OEE measures the effectiveness of TPM activity.

In our example, we have assumed an eight-hour shift, or 480 minutes. The actual time planned by management is 435 minutes (30 minutes for lunch break and two 15-minute breaks). Note that if there were one or more tool changeovers or setups, then this time should be deducted from the 480 minutes of total shift time.

In our case, the available time is 435 minutes. Therefore, the availability rate is = 435 minutes/480 minutes = 90.6%

The performance rate = actual output/target output.

Assuming our equipment is rated to produce 120 parts per minute, our target output = 435 × 120 = 52,200 parts.

To calculate the actual output, we need to know the time lost during the normal working of the shift apart from the time allowed by the management, i.e., 45 minutes.

Assume that we lost the following times:

Idle time	10 minutes
Setup time	25 minutes
Maintenance time	20 minutes
Miscellaneous time lost	10 minutes
Total	65 minutes

Therefore, the actual output will be (435 minutes − 65 minutes lost) × 120 = 44,400 parts

Actual output = 44,400 parts

$$\text{The performance rate} = \text{actual output/target output} = 44,400/52,200 = 85\%$$

$$\text{Quality rate} = \text{good output/actual output}$$

Assuming our parts went out of specification for one hour, so we scrapped 120 × 60 = 7200.

Also, we reworked 2000 parts produced during the setup time.

Now, quality rate = (44,400 − 7200 − 2000) parts/44,400 = 79.2%.

Calculate the OEE and analyze the result:

$$\text{OEE} = 90.6\% \text{ (availability rate)} \times 85\% \text{ (performance rate)} \times 79.2\% \text{ (quality rate)}$$

$$\text{OEE} = 0.906 \times 0.85 \times 0.792 = 0.61 \text{ or } 61\%$$

Analyze Results of OEE Daily

As we can see, the above OEE result is for one shift. We also have the data of the lost time and the wastes. Most of the downtime can be reduced or eliminated through root cause analysis techniques explained earlier and subsequently in the sister book *The Tactical Guide to Six Sigma Implementation*.

The following facts are worth knowing to understand the effect of OEE:

- Initial stage of TPM implementation yields OEE levels less than 50%
- World-class TPM companies can expect an OEE of 85%
- Typically, breakdown is ten times more expensive than the TPM program (Figure 8.47)

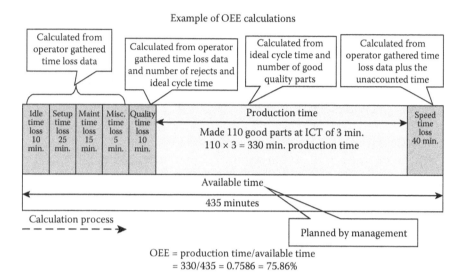

Example of OEE calculations

FIGURE 8.47
Example of OEE calculations.

STANDARD WORK

Let me tell you about the success story of Southwest Airlines. It is the largest low-cost carrier in the United States, and is headquartered in Dallas, Texas. It was started by Rollin King and Herb Kelleher in the early 1970s.

Their initial goal was to provide air service between two cities at a cost less than it would take you to drive between two cities. They bought four Boeing 737 planes. Three planes were always used and the fourth one was used as a spare. They started the service between San Antonio, Dallas, and Houston—the three major cities of Texas, situated in a triangle, and each one separated by approximately 350 km.

Their core competency is in standardization. Southwest Airlines' standardization falls into four areas:

- A standard product—open seating at equal prices. No classes. You sit anywhere you find an empty seat.
- A standard process—in and out of a gate in less than 20 minutes
- A standard part—every plane in their fleet is a Boeing 737
- A standard training—standard operating procedures promote interchangeable and multifunctional staff. Absenteeism has a minimal effect.

Since its inception in the 1970s, Southwest has suffered made a loss! Today, Southwest has more than 46,000 employees and operates more than 3400 flights per day. As of August 2012, Southwest Airlines operates scheduled services to 77 destinations in 40 states.

This is the power of standardization.

Standard Work Definition

It is a documented system in which production workers develop and follow a repeatable sequence of tasks within a work assignment. The standardized work sequence represents the best practices for the operator to follow in the completion of his/her job.

What is the benefit of standard work?

- Documented process
- Improved efficiency because of process stability. Stability means repeatability
- Improved training procedures, resulting in organizational learning
- Work balancing
- Better utilization of equipment, tools, and persons
- Basis of all Lean activities

What are the resource requirements?

- Operators and support areas documenting process and working out takt time, cycle time, and line balancing, etc.
- Approximately three to five days including four hours of training. Training documentation can then be made available for future training

What are the elements and steps? This is composed of three elements:

1. Takt time
2. Cycle time
3. Standard work sequence
4. Standard work-in-process

Takt Time

Takt is German for the baton that an orchestra conductor uses to regulate the speed, beat, or rhythm at which musicians play. In a Lean environment, we can say that it is the rhythm with which a product or service provider should pace the activity or work to meet customer demand.

Rules of Takt Time

- It is the time required between the completion of successive units of product or service
- Takt time is calculated for each process step and is used to pace the work
- Takt time becomes extremely difficult to manage when the customer—demand environment is unstable
- Takt time must be reviewed and updated with changes in demand
- Takt time (in minutes) = available time per period/required customer demand in that period
- Assuming net available time of 420 minutes per shift and the demand of 210 gadgets per shift
- Takt time = 420/210 = two minutes per gadget
- Here, the flow rate of the gadget production steps must be less than or equal to two minutes per gadget
- If it is more than two minutes, a backlog will be developed with respect to customer demand, causing late deliveries and, in the long run, the customer will find another supplier!

Cycle Time

Unlike takt time, which is related to customer demand and can be called demand rate, the cycle time is the designed time to complete one unit according to the standard work sequence, and can be called production time or work sequence time.

Standard Work Sequence

Process Instruction Document

The process instruction document for standard work should have a precise and detailed description of each work activity, with visual instructions written from the perspective of the operator that details the sequence,

methods, and tools required and that are critical to the quality characteristics of an operation.

The standard work is created by means of the following charts:

- Time observation chart
- Process capacity chart
- Operator balance hart
- Standardized work chart
- Standardized work combination table

Time Observation Chart

Time observation chart for a typical assembly operation is shown in Figure
8.48. It is self explanatory.

Process Capacity Chart

This chart determines the production capacity based on the following calculations (Figure 8.49):

1. Find working hours per shift. For example, 480 minutes = 2880
   seconds
2. Find process time per piece in seconds = manual work + machine
   work. For example, 5 + 40 = 45 seconds
3. Find setup time. For example, 150 seconds
4. The production lost during setup time = 300
5. Extra time per piece needed to compensate for setup time =
   150/300 = 0.5 seconds
6. Total time per piece = 45 + 0.5 = 45.5 seconds
7. Capacity per shift = 2880/45.5 = 633 pieces

Operator Balance Chart

See Figure 8.50.

Standardized Work Chart

See Figure 8.51.

Standardized Work Combination Table

See Figure 8.52.

Model: ABC-6992
Desciption: 18 pcs
Part no.
Process: Assembly

Operator: _____ Various
Observer: _____ Brenda

Date: 10-07-2009

Task	1	2	3	4	5	6	7	8	9	10	Task time secs	Task time 1.15	Task time min	Comments
Select Blocks	850										850	54	0.91	Incidental work
Check drawing for latest rev.	443										443	28	0.47	Incidental work
Place blocks on trolly	131										131	8	0.14	Incidental work
Move to cell	60										60	4	0.06	Incidental work
Review work instruction, tools, print and materials	714										714	821	1.58	Incidental work
Place parts in their containers	315										315	82	1.37	Incidental work
Obtain model labels	108										108	7	0.12	LEADER
Place the block in a fixture	120	106	98	132	115	122	136	121	134	156	124	143	2.38	Sub assembly 1
Assemble plugs	70	61	75	78	78	68	64	81	69	70	71	82	1.37	
Stamp model number	29	20	17	27	32	26	27	25	33	19	26	29	0.49	Sub assembly 2
Fix labels	22	21	18	24	19	23	22	18	24	22	21	24	0.41	
Assemble expander plugs	151	121	106	110	115	123	142	136	123	132	126	145	2.41	
Assemble check valve	58	65	69	72	63	60	70	59	67	57	64	74	1.23	
Stamp valve	85	82	79	88	76	89	80	84	88	79	83	95	1.59	Sub assembly 3
Assemble expander plugs	28	25	32	29	24	33	35	28	27	27	29	33	0.55	
Assemble dual piston	89	97	91	97	90	92	96	94	89	90	93	106	1.77	
Assemble check valve	45	49	52	49	54	50	48	55	51	53	51	58	0.97	
Assemble expander plugs	39	41	44	40	43	45	39	38	42	43	41	48	0.79	Sub assembly 4
Assemble DX valve	256	266	245	253	264	269	250	268	249	256	258	296	4.94	
Assemble connectors	196	279	205	235	241	238	260	259	232	243	239	275	4.58	
Assemble coil	293	231	289	234	286	271	265	289	269	273	271	311	5.18	
Remove from fixture	80	99	90	87	95	84	88	91	86	85	89	102	1.70	
Set-up for test	600										600	690	11.50	
Test	427	390	432	385	435						414	476	7.93	
Pack	451	464	475	432	426						450	517	8.62	
Cycle time	12	11	11	12	12	1000	1009	1007	996	1004	3607	2419	40	

FIGURE 8.48

Time observation chart.

Part name: Stud											Dept.			101	Approved
Part number: 12345											Shift hrs			480 mins.	

PROCESS CAPACITY CHART

Process no.	Process name	Machine no.	Basic time secs				Set up			Per piece	Process capacity (pcs per shift)
			Manual work	Auto cycle	Completion time	Quantity during exchange	Tool exchange time	Exchange time per piece	Total	
1	Form head	1	5.0	40.0	45.0	300.0	150.0	0.5	45.5	633.0
2	Thread	2	6.0	15.0	21.0	500.0	120.0	0.2	21.2	1358.0
3	Deburr	Bench	30.0		30.0				30.0	960
4										
5										
6										

Note: 633 = (480/45.5) × 60

FIGURE 8.49

Production capacity calculation chart.

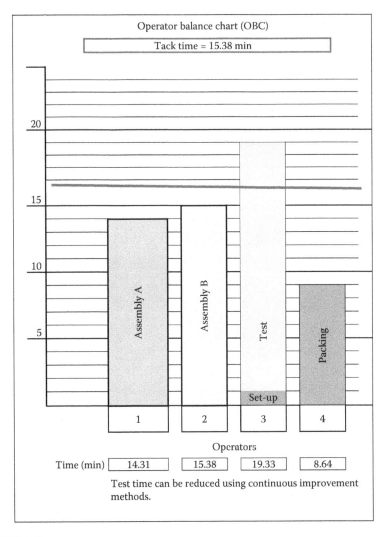

FIGURE 8.50
Operator balance chart.

Standard Work-in-Process

Standard work-in-process is the determined minimum amount of material needed to keep the Standard Work flowing. In order to do this Lean Tools like U-shaped manufacturing cell, Continuous flow (Kanban), and One Piece Flow are used.

Standard work chart propuesta

Part name: MCD-6494	Part #	18	Number of operations in cell:	4	Operator number:	4	Cell: MCD	From: Ensamble	To: Empaque	Date: 11-Julio-07
Process name	Ensamble									
sequence	Activation									
1	Colocar pieza en el fixture									
	Colocar sae plug									
	Estampado									
	Etiquetas									
	Colocar expander									
	Colocar check valve									
	Estampados									
	Colocar expander									
	Colocar dual piston									
	Colocar check valve									
	Colocar expander									
2	Colocar DG									
	Colocar conectores									
	Acomodar coil									
	Quitar la pieza									
3	Setup									
4	Test									
5	Empaque									

Quality check	◆	Safety	✚	Std. WIP	◯	# Parts WIP	1	Line rate	Cycle time 855 seg	Takt time 1781 seg

015
Oil gear

Standardized work combination sheet

| Part number | 98-097-654 | | | Date prepared | March-10-05 | Work area | Manifold assembly | Daily demand | 231 units |
| Part name | HT injection manifold | | | | | | Takt time | | 117 second |

| Work sequence | | Time | | | Operation time (seconds) |
Step	Operation	Man	Auto	Walk	
1	Obtain manifold from conveyor	15		5	
2	Place in wash tub	10		1	
3	Press button to start wash tub	1			
4	Wash tub cycle		10		
5	Obtain manifold from wash tub	6			
6	Place manifold on work bench	2		5	
7	Obtain 4 valve from stock area	2		5	
8	Remove packaging from valves	10			
9	Place valves on manifold	8			
10	Obtain screws and hand tighten	7			
11	Obtain torque wrench and torque to 50 ft lbs	3			
12	Record torque values on tracker	8			
13	Mask off valve assembly	9			
14	Aside to pallet	4		5	
15	Move to conveyor to start next assembly				
		95	10	21	

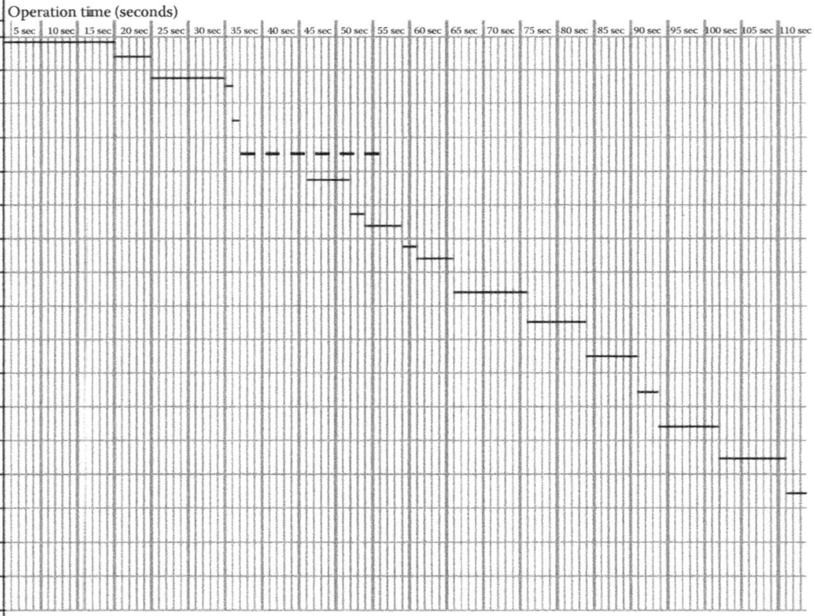

FIGURE 8.52
Standardized work combination table.

CONTINUOUS IMPROVEMENT

Standardized work is the foundation of continuous improvement.

> Today's standardization is the necessary foundation on which tomorrow's improvement will be based. If you think of 'standardization' as the best you know today, but which is to be improved tomorrow—you get somewhere. But if you think of standards as confining, then progress stops.

> **Henry Ford (1926)**

> It is impossible to improve any process until it is standardized. If the process is shifting from here to there, then any improvement will just be one more variation that is occasionally used and mostly ignored. One must standardize, and thus stabilize the process, before continuous improvement can be made.

> **Masaki Imai, Kaizen, 1986**

Standardized Work Sustains Continuous Improvements

Kaizen means steady but continual improvement. Dr. Russ Ackoff explains the difference between "continuous improvement" and "discontinuous improvement" as seen through the lens of systems thinking.

One of the most difficult aspects of introducing and implementing a continuous improvement system is assuring its continuity.

When a company introduces continuous improvement, it experiences some initial success, but soon such success disappears like scattered clouds on a rainy season and after a while nothing is left, and management keeps looking for a new flavor of the month.

There is a study concluding that two-thirds of managers who started programs in continual improvement said the programs failed. The study also concluded that they failed because they were not embedded in systems thinking. Dr. Ackoff describes systems thinking in his characteristic profound, insightful, and funny way, reminding us that only the system can perform its functions. None of its parts can. The products and services an organization provides and the functions it performs are all the "product of their interactions." Setting out to improve the parts independently will not improve the whole system.

My mind went immediately to sports. I remember when on March 17, 2012—Mirpur Bangladesh, the most-awaited 100th international century by Sachin Tendulkar went in vain as India suffered a humiliating loss by five wickets in their Asia Cup encounter. One part of the system (Sachin as a batsman) excelled, but the team (system) failed because it did not perform in cohesion toward a goal.

Kaizen

Kaizen is a state of mind or philosophy in which one seeks to continuously improve—never resting or accepting the status quo.

Kai = to take apart
Zen = to make good
Kaizen = continuous improvement

The seven conditions for successful implementation of kaizen strategy are

- Top management commitment
- Top management commitment
- Top management commitment
- Setting up an organization dedicated to promote kaizen
- Appointing the best available personnel to manage the kaizen process
- Conducting training and education
- Establishing a step-by-step process for kaizen introduction

All conditions are important. Without top management supporting every move, however, the trial will be short-lived regardless of other preconditions.

Top management may express commitment in many different ways, and it must take every opportunity to preach the message, become personally involved in following up the progress of kaizen, and allocate resources for successful implementation.

More on kaizen follows later in this section.

Kaikaku

Kaikaku means revolution, or radical improvement. Both kaizen and kaikaku are needed for successful continuous improvement. The following 10 kaikaku commandments are all good basic principles to start any improvement journey. It is a top-down initiative to activate a bottom-up empowerment for change.

10 Kaikaku Commandments by Hiroyuki Hirano

- Throw out the traditional concept of manufacturing methods. Think outside the "box"
- Think about how the new method will work, not how it won't work
- Don't accept excuses; totally deny the status quo
- Don't seek perfection; a 50% implementation rate is fine as long as it's done on the spot
- Correct mistakes the moment they are found
- Don't spend money on kaikaku
- Problems give you a chance to use your brains
- Ask "Why?" five times
- Ten person's ideas are better than one person's knowledge
- Kaikaku knows no limits

Figure 8.53 explains kaizen and kaikaku pictorially.

Continuous improvement tools

1. PDCA
2. Organizing kaizens for improvement
3. Kaizen
4. A3 reports

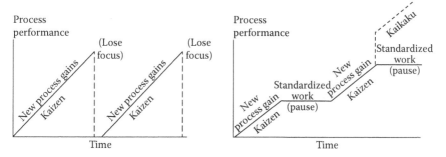

Without standardization, there cannot be improvements.

Kaizen (continuous "incremental" improvements)
Kaikaku (radical "breakthrough" improvements)

FIGURE 8.53
Standardized work sustains improvement.

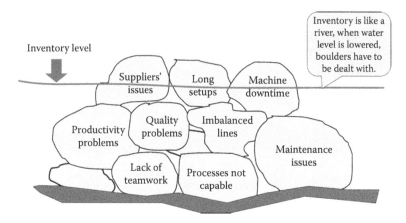

FIGURE 8.54
Inventory buildup and excess production hide many "sins."

The first thing to do is to kill the monster called overproduction or built up inventory. Reducing waste from overproduction (inventory) brings problems to the surface and forces their resolution (Figure 8.54).

PDCA—Problem Solving and Process Improvement Model

PDCA sounds simple and is easily glossed over, but if well done, it is great for problem-solving and process improvement (Figure 8.55). PDCA is considered a vital element of the Toyota Production System.

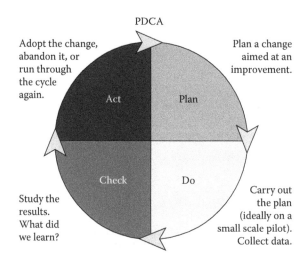

FIGURE 8.55
PDCA improvement cycle.

PDCA Improvement Cycle

Plan—Select the process/problem to work on. Describe the current process. Collect data: 4W1H. With a problem, take time to define the problem and agree on root causes (use fishbone, five whys, Pareto, etc.). Develop a solution and action plan.

Do—Implement the process change or problem solution. Start on a small scale (pilot).

Check—Is the process change working as planned? Is the problem solved? If not, why not, and what can we learn?

Act (or Standardize)—This is a vital, but frequently neglected, step. If you were successful, standardize the solution. Celebrate and congratulate. Otherwise, continue in the cycle to plan for further improvements.

Tools of Problem Solving That Are Applied as Part of PDCA

Cause and Effect Diagram

A fishbone diagram, also called cause-and-effect diagram, or Ishikawa diagram (Figure 8.56). The fishbone diagram identifies many possible causes for an effect or problem. It can be used to structure a brainstorming session. It immediately sorts ideas into useful categories.

When to Use a Fishbone Diagram

- When identifying possible causes for a problem
- Especially when a team's thinking tends to fall into ruts
- Fishbone diagram procedure
- Materials needed: flipchart or whiteboard, marking pens

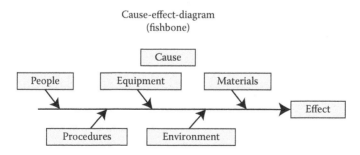

FIGURE 8.56
Cause and effect diagram.

- Agree on a problem statement (effect). Write it at the center right of the flipchart or whiteboard. Draw a box around it and draw a horizontal arrow running to it
- Brainstorm the major categories of causes of the problem. If this is difficult, use generic headings
  - Methods
  - Machines (equipment)
  - People (manpower)
  - Materials
  - Measurement
  - Environment

How to Start?

Write the categories of causes as branches from the main arrow. Brainstorm all the possible causes of the problem. Ask, "Why does this happen?" As each idea is given, the facilitator writes it as a branch from the appropriate category. Causes can be written in several places if they relate to several categories.

Again ask, "Why does this happen?" about each cause. Write subcauses branching off the causes. Continue to ask "Why?" and generate deeper levels of causes. Layers of branches indicate causal relationships.

When the group runs out of ideas, focus attention to places on the chart where ideas are few.

Fishbone Diagram Example

This fishbone diagram was drawn by a manufacturing team to try to understand the source of periodic iron contamination. The team used the six generic headings to prompt ideas. Layers of branches show thorough thinking about the causes of the problem (Figure 8.57).

Pareto Chart

The Pareto principle deals with the vital few and trivial many rule. It is also called the "80:20 Rule"

It is named after Vilfredo Pareto—an Italian economist (Figure 8.58).

- He observed in 1906 that 20% of the Italian population owned 80% of Italy's wealth
- He then noticed that 20% of the pea pods in his garden accounted for 80% of his pea crop each year

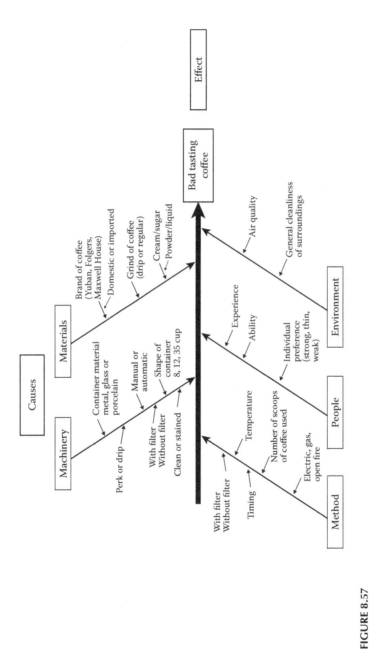

FIGURE 8.57
Fishbone diagram (Ishikawa diagram).

FIGURE 8.58
Vilfredo Pareto.

The Pareto Principle

- A small number of causes is responsible for a large percentage of the effect—usually a 20% to 80% ratio.
- This basic principle translates well into quality problems—most quality problems result from a small number of causes.
- You can apply this ratio to almost anything, from the science of management to the physical world.
- Addressing the most troublesome 20% of the problems will solve 80% of it.
- Within your process, 20% of the individuals will cause 80% of your headaches.
- Of all the solutions you identify, about 20% are likely to remain viable after adequate analysis.
- 80% of the work is usually done by 20% of the people.

A Pareto chart is a useful tool for depicting these and other relationships graphically. It is a simple histogram style graph that ranks problems in order of magnitude to determine the priorities for improvement activities.

The goal is to target the largest potential improvement area, then move on to the next, then the next, and in so doing address the area of most benefit first. The chart can help show you where allocating time, human, and financial resources will yield the best results.

While the rule is not an absolute, one should use it as a guide and reference point to ask whether or not you are truly focusing on 20% (the vital few) or 80% (the trivial many). True progress results from a consistent focus on the 20% most critical objectives.

The simplicity of the Pareto concept makes it prone to being underestimated and overlooked as a key tool for quality improvement. Generally, individuals tend to think that they know the important problem areas requiring attention… if they really knew, why do problem areas still exist?

Although the idea is quite simple, to gain a working knowledge of the Pareto principle and its application, it is necessary to understand the following basic elements:

Pareto Analysis of Printing Defects

The Pareto diagram is a combined bar chart and line diagram based on cumulative percentages.

Pareto Diagram of Total Printing Defects

Similar Pareto diagrams can be prepared for each press. We can observe here that color variation is the major defect among all defects and the press 22 × 28 is the major contributor to this defect (Figure 8.59).

Knowing this, we can apply the PDCA and RCA (root cause analysis) method to reduce this defect to a minimum. Actions can then be taken for the next highest defect (Figure 8.60).

Defect	Press 22 × 28	Press 38" 2-C	Press 77" 5-C	Press 77" 4-C	Total Defects
Color Variation	550	430	234	476	1690
Misregister	150	27	31	265	473
Lint/dirt etc.	50	45	80	10	185
Scuffing	10	14	3	60	87
Excess spray	16	21	30	5	72
Other	30	37	21	30	118
Totals	806	574	399	846	2625
% Waste	30.7	21.87	15.2	32.23	100

FIGURE 8.59
Pareto analysis of printing defect.

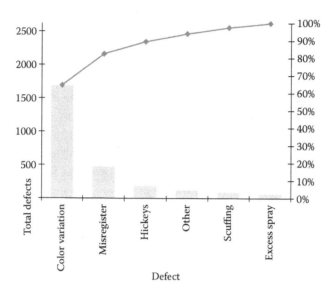

FIGURE 8.60
Pareto chart for total printing defects.

Check Sheet

Also called *defect concentration diagram.* A check sheet is a structured, prepared form for collecting and analyzing data. This is a generic tool that can be adapted to a wide variety of purposes.

When to Use a Check Sheet

- When data can be observed and collected repeatedly by the same person or at the same location
- When collecting data on the frequency or patterns of events, problems, defects, defect location, defect causes, etc.
- When collecting data from a production process

Check Sheet Procedure

- Decide what event or problem will be observed. Develop operational definitions
- Decide when data will be collected and for how long
- Design the form. Set it up so that data can be recorded simply by making check marks or Xs or similar symbols and so that data do not have to be recopied for analysis
- Label all spaces on the form

- Test the check sheet for a short trial period to be sure it collects the appropriate data and is easy to use
- Each time the targeted event or problem occurs, record data on the check sheet

Check Sheet Example

Figure 8.61 shows a check sheet used to collect data on telephone interruptions. The tick marks were added as data were collected over several weeks.

Control Chart

Figure 8.62 shows an example of a control chart.

Histogram

A frequency distribution shows how often each different value in a set of data occurs. A histogram is the most commonly used graph to show frequency distributions. It looks very much like a bar chart, but there are important differences between them.

Telephone interruptions

| Reason | Day | | | | | |
	Mon	Tues	Wed	Thurs	Fri	Total
Wrong number	ЖЖ	II	I	ЖЖ	ЖЖ II	20
Info request	II	II	II	II	II	10
Boss	ЖЖ	II	ЖЖ II	I	IIII	19
Total	12	6	10	8	13	49

FIGURE 8.61
Check sheet for telephone interruptions.

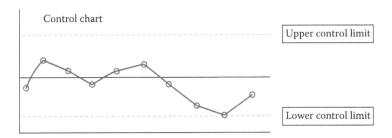

FIGURE 8.62
Control chart.

When to Use a Histogram

- When the data are numerical
- When you want to see the shape of the data's distribution, especially when determining whether the output of a process is distributed approximately normally
- When analyzing whether a process can meet the customer's requirements
- When analyzing what the output from a supplier's process looks like
- When seeing whether a process change has occurred from one period to another
- When determining whether the outputs of two or more processes are different
- When you wish to communicate the distribution of data quickly and easily to others

Histogram Construction

Collect at least 50 consecutive data points from a process. Use the histogram worksheet (below) to set up the histogram. It will help you determine the number of bars, the range of numbers that go into each bar, and the labels for the bar edges. After calculating bar width in step 2 of the worksheet, use your judgment to adjust it to a convenient number. For example, you might decide to round 0.9 to an even 1.0. The value for *W* must not have more decimal places than the numbers you will be graphing.

Step 1: Determine the number of bars using the following table:

Number of Data Points	Number of Bars *B*
50	7
60	8
75	9
100	10
125	11
150	12
200	14

Step 2: Determine the width of the bar (upper edge – lower edge of each bar) *W*

$$W = \text{range of the data points/number of bars B}$$

Example:

Number of data points = 100
Lowest data point: 109.25
Largest data point: 141.88
Number of bars = 10
Range = 32.6
Width W = 3.26

Add 3.26 to the lowest data point (109.25), go to the next bar, etc.

Histogram Analysis

Before drawing any conclusions from your histogram (Figure 8.63), satisfy yourself that the process was operating normally during the period being studied. If any unusual events affected the process during the period of the histogram, your analysis of the histogram shape probably cannot be generalized to all periods (Figure 8.64).

Scatter Diagram

Also called scatter plot or *X–Y* graph. The scatter diagram graphs pairs of numerical data, with one variable on each axis, to look for a relationship

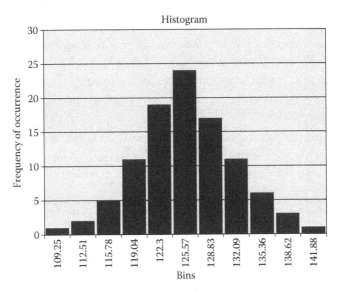

FIGURE 8.63
Histogram.

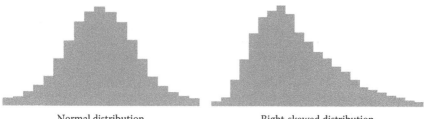

Normal distribution   Right-skewed distribution

FIGURE 8.64
Typical histogram shapes.

between them. If the variables are correlated, the points will fall along a line or curve. The better the correlation, the tighter the points will hug the line.

When to Use a Scatter Diagram

- When you have paired numerical data
- When your dependent variable may have multiple values for each value of your independent variable
- When trying to determine whether the two variables are related, such as…
- When trying to identify potential root causes of problems
- After brainstorming causes and effects using a fishbone diagram, to determine objectively whether a particular cause and effect are related
- When determining whether two effects that appear to be related both occur with the same cause
- When testing for autocorrelation before constructing a control chart

Scatter Diagram Procedure

- Collect pairs of data where a relationship is suspected
- Draw a graph with the independent variable (X) on the horizontal axis and the dependent variable (Y) on the vertical axis. For each pair of data, put a dot or a symbol where the X-axis value intersects the Y-axis value (if two dots fall together, put them side by side, touching, so that you can see both)
- Look at the pattern of points to see if a relationship is obvious. If the data clearly forms a line or a curve, you may stop. The variables are correlated. You may wish to use regression or correlation analysis now. Otherwise, complete steps 4 through 7

- Divide points on the graph into four quadrants. If there are *X* points on the graph
  - Count *X*/2 points from top to bottom and draw a horizontal line
  - Count *X*/2 points from left to right and draw a vertical line
  - If the number of points is odd, draw a line through the middle point
  - Count the points in each quadrant. Do not count points on a line
  - Add the diagonally opposite quadrants. Find the smaller sum and the total of points in all quadrants
  - *A* = points in upper left + points in the lower right
  - *B* = points in upper right + points in the lower left
  - *Q* = the smaller of *A* and *B*
  - *N* = *A* + *B*
  - Look up the limit for *N* on the trend test table (Table 8.8)
  - If *Q* is less than the limit, the two variables are related
  - If *Q* is greater than or equal to the limit, the pattern could have occurred from random chance

TABLE 8.8

Trend Test Table

N	Limit	N	Limit
1–8	0	51–53	18
9–11	1	54–55	19
12–14	2	56–57	20
15–16	3	58–60	21
17–19	4	61–62	22
20–22	5	63–64	23
23–24	6	65–66	24
25–27	7	67–69	25
28–29	8	70–71	26
30–32	9	72–73	27
33–34	10	74–76	28
35–36	11	77–78	29
37–39	12	79–80	30
40–41	13	81–82	31
42–43	14	83–85	32
44–46	15	86–87	33
47–48	16	88–90	34
49–50	17	90	35

Scatter Diagram Example

The manufacturing team suspects a relationship between product purity (percent purity) and the amount of iron (measured in parts per million or ppm). Purity and iron are plotted against each other as a scatter diagram, as shown in Figure 8.65.

There are 24 data points. Median lines are drawn so that 12 points fall on each side for both percent purity and ppm iron.

To test for a relationship, they calculate:

A = points in upper left + points in lower right = 9 + 9 = 18
B = points in upper right + points in lower left = 3 + 3 = 6
Q = the smaller of A and B = the smaller of 18 and 6 = 6
N = A + B = 18 + 6 = 24

Then, they look up the limit for N on the trend test table. For N = 24, the limit is 6. Q is equal to the limit. Therefore, the pattern could have occurred from random chance, and no relationship is demonstrated.

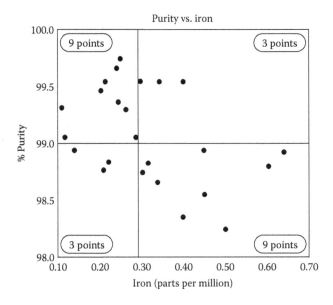

FIGURE 8.65
Iron purity scatter data plot.

Scatter Diagram Considerations

Here are some examples of situations in which you might use a scatter diagram:

Variable *A* is the temperature of a reaction after 15 minutes. Variable *B* measures the color of the product. You suspect that higher temperatures make the product darker. Plot the temperature and color on a scatter diagram.

Variable *A* is the number of employees trained on new software, and variable *B* is the number of calls made to the computer help line. You suspect that more training reduces the number of calls. Plot the number of people trained versus the number of calls.

To test for autocorrelation of a measurement being monitored on a control chart, plot this pair of variables: Variable *A* is the measurement at a given time. Variable *B* is the same measurement, but at a previous time. If the scatter diagram shows correlation, do another diagram where variable *B* is the measurement two times previously. Keep increasing the separation between the two times until the scatter diagram shows no correlation.

Even if the scatter diagram shows a relationship, do not assume that one variable caused the other. Both may be influenced by a third variable. When the data are plotted, the more the diagram resembles a straight line, the stronger the relationship.

If a line is not clear, statistics (*N* and *Q*) determine whether there is reasonable certainty that a relationship exists. If the statistics say that no relationship exists, the pattern could have occurred by random chance. If the scatter diagram shows no relationship between the variables, consider whether the data might be stratified.

If the diagram shows no relationship, consider whether the independent (*X*-axis) variable has been varied widely. Sometimes a relationship is not apparent because the data do not cover a wide enough range. Think creatively about how to use scatter diagrams to discover a root cause. Drawing a scatter diagram is the first step in looking for a relationship between variables.

Flowchart

Description

Flowcharts are used in designing and documenting processes. Like other types of diagrams, they help the viewer visualize and understand a process and potentially discover flaws, bottlenecks, and other less-obvious issues within it.

There are many different types of flowcharts, each with its own collection of boxes and notational conventions. The two most common types of boxes in a flowchart are a processing step, usually called activity and denoted as a rectangular box, and a decision, usually denoted as a diamond.

A flowchart is described as "cross-functional" when the page is divided into "swim lanes" describing the control of different organizational units. A symbol appearing in a particular "lane" is within the control of that organizational unit. This technique allows the author to locate the responsibility for performing an action or making a decision correctly, showing the responsibility of each organizational unit for different parts of a single process.

Benefits

- Visual representation of the process or issue under study
- Easily created using simple applications or by hand
- Involves teamwork and uses people knowledge to document the flowchart
- Applicable to all aspects of an organization

How to Use

Step 1. Assemble a team of people with knowledge of the process. The scope must be defined prior to starting the mapping: Where does the process start and where does it finish?

Step 2. Starting at the beginning of the process, document the steps in sequence from start to finish.

Step 3. Add all the different loops that are within or between the different processes.

Step 4. Connect the boxes (activities) and decision points.

Step 5. Validate the final map with the team and management.

Relevant Definitions

Common alternate names include: flowchart, process flowchart, functional flowchart, process map, process chart, functional process chart, business process model, process model, process flow diagram, work flow diagram, and business flow diagram.

Example   An organization specializing in printing multicolor brochures for small businesses has realized that as the number of orders increases, its lead time starts to creep upwards and promised delivery dates are not met on a regular basis. To understand the process and identify actions to fix the issue, an improvement team decides to map the process. Because several departments are involved, the team uses swim lanes. The map below shows the different activities and the responsible departments. The process map makes it obvious that there is too much back and forth between the sales and graphic design departments, causing huge delays (Figure 8.66).

Validate Root Cause of a Problem

Through experimentation, observation (or simulation), verify that you have found the true root cause(s). See if you can generate and eliminate the problem by repetitively installing and removing the cause (Figure 8.67).

When Problem Solving Is Not Enough, Deploy DMAIC (Six Sigma)

The DMAIC process is shown in Figure 8.68.

Employee Involvement

A cartoon depicting employee involvement is shown in Figure 8.69 (Table 8.9). G. Truell on the "involvement of people":

> Organizations must provide effective responses to increasing quality expectations for goods and services through employee involvement (EI). EI is a vital component of any effort to develop an organization as a proactive system that responds to changes rapidly in the face of dramatic shifts in national and international marketplaces.

As supervisors begin to decide when to involve team members in problem solving and decision making, they frequently ask themselves the following questions:

- When should I involve others?
- Who should be involved?
- To what extent should I involve them?
- Who should make the final decision?

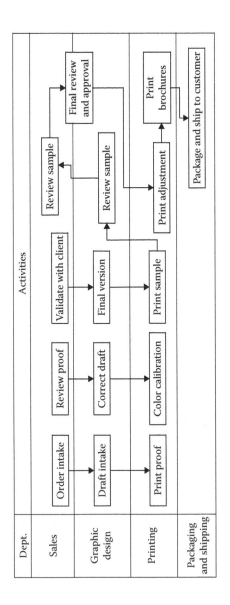

FIGURE 8.66
Flowchart.

Turn it on, turn if off

FIGURE 8.67
Criteria to validate the root cause of a problem.

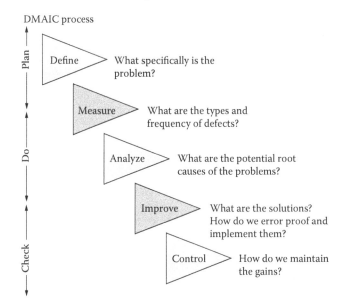

FIGURE 8.68
DMAIC process for continuous improvement.

There are three key factors in problem solving and decision making involving people (Table 8.9).

- Possession of data—amount, kind, and quality of data the supervisor possesses
- Acceptance by others—degree to which support, understanding, and commitment of others is important in implementing the decision
- Time—amount of time the supervisor has before a decision must be made and implemented

It is important to understand how the final decision will be made, especially when using the consult phase, which involves thinking that you get

FIGURE 8.69

Cartoon about failed communication regarding employee safety risk.

TABLE 8.9

Employee Involvement Stages

Type of Organization	Ways of Employee Involvement
Traditional organization	Manager solves problems
Organization having quality circles	Employees help identify and solve problems
Organization where problem solving is a part of employee's job	Self-managing work teams

input but you make a decision, and the join phase where the members of the team make a decision (Figure 8.70).

KAIZEN

Guidelines for Facilitating a Successful Kaizen Event

Prepare for the Event, Team Roles, and Issues

Kaizen Team Members and their Roles Explained (Figure 8.71):

- Timekeeper/checker
  - Keeps team meetings on time schedule and focused on agenda.

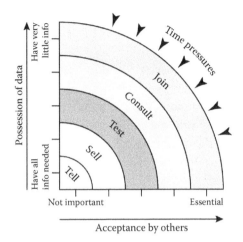

FIGURE 8.70
G. Truell's three factors regarding the involvement of people.

Kaizen team roles

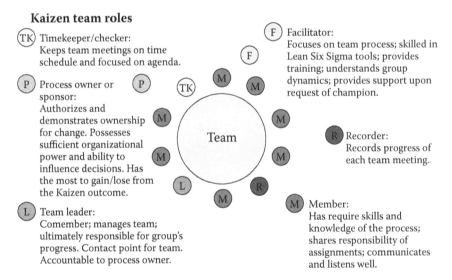

(TK) Timekeeper/checker:
Keeps team meetings on time
schedule and focused on agenda.

(P) Process owner or
sponsor:
Authorizes and
demonstrates ownership
for change. Possesses
sufficient organizational
power and ability to
influence decisions. Has
the most to gain/lose from
the Kaizen outcome.

(L) Team leader:
Comember; manages team;
ultimately responsible for group's
progress. Contact point for team.
Accountable to process owner.

(F) Facilitator:
Focuses on team process; skilled in
Lean Six Sigma tools; provides
training; understands group
dynamics; provides support upon
request of champion.

(R) Recorder:
Records progress of
each team meeting.

(M) Member:
Has require skills and
knowledge of the process;
shares responsibility of
assignments; communicates
and listens well.

FIGURE 8.71
Kaizen team members.

- Process owner or sponsor
  - Authorizes and demonstrates ownership for change. Possesses sufficient organizational power and ability to influence decisions. Has the most to gain/lose from the kaizen outcome.
- Team leader
  - Comember; manages team. Ultimately responsible for group's progress. Contact point for team. The team leader is accountable

to the process owner and does not supervise the team or apply his/her own solutions, but leads the team to its goals.
- Facilitator (Lean coordinator)
  - Focuses on team process; skilled in Lean Six Sigma tools; provides training; understands group dynamics: provides support upon request of champion. Webster's defines "facilitate" as "to make easier." His job is to make the team's work easier by managing the process and the dynamics. As a facilitator, he is expected to guide, assist, and coach the team (not participate in the work of the team).
- Recorder
  - Records progress of each team meeting.
- Team member
  - Has the required skills and knowledge of the process; shares responsibility of assignments; communicates and listens well.

Team Members' Roles

- Represent all major stakeholder groups involved at various stages of the process (including upstream and downstream processes). Select team members by asking for volunteers.
- Possess authority and skills to gather data, conduct problem solving, recommend changes, and implement solutions.
- View issues/problems as a priority for improvement and will see it through to its natural end.
- Represent the voice of the customer.
- Willing to put team goals ahead of individual goals.
- Able and willing to attend all team meetings.
- Open to change, i.e., be willing to look outside the boundaries of their current job. Team members should be willing and able to take risks and to look at things in a new way.

Team Size

Typical kaizen teams include five to ten people who often represent different cross-functional areas. Team composition is usually divided into thirds:

- One-third are the very hands-on "process experts" (stakeholders who will be affected by the outcome)
- One-third are management "decision-makers" who have authority to act (either as a full-time member or through their sponsorship)
- One-third are outsiders, people from direct upstream and downstream processes who can look at the process with fresh eyes

Three words about teams: communications, communications, communications!

Communication links increase as the size of the group increases by

$$n [n - 1]/2 \text{ (where } n = \text{number of people).}$$

Therefore, when $n = 5$, $5[5 - 1]/2 = 10$ links
When $n = 20$, $20[20 - 1]/2 = 190$ links

The larger the group is, the more communication traffic there is to handle. The ideal size limit for a group is five to ten members.

- Sponsor (also called value stream manager, who is not a team member):
  - Clearly articulates the business case or rationale for why this change is needed. Establishes a sense of urgency.
  - Communicates strong ownership and personal commitment for this change. Accepts some degree of risk.
  - Establishes priorities for improvement in a way which ensures that the organization's strategy is achieved.
  - Creates (when needed) a guiding coalition of people with enough power to lead the change and work together as a team.
  - Ensures adequate resources are available, i.e., time, money, people, authority to act, access to decision makers.
  - Aligns the reward and recognition systems.

Team Charter (Scope)

A key part of a successful kaizen event is that each team member and team leader can answer the following questions (Figure 8.72):

- What is the driving issue and why has this process been selected? (Why are we here?)
- What are the specific metrics we are to improve?
- What is our role and that of others?
- What are the deliverables? (Design manufacturing cell, re-layout area, etc.)
- What, if any, are the constraints? What are off-limits?
- What is the deadline? How much time are we expected to spend?
- What happens to our "regular jobs" while we are involved in this event?

This brings us to "kaizen event issues."

Kaizen events issues: Conducting Kaizens without a purpose

Kaizen events have their places. Kaizen create little islands of improved flow at the process level, but you need to link those islands together in a system-level flow.

The future state Value C stream Map "vision" is used as a blueprint to ensure that individual activities, including Kaizens, will ultimately fit together. Don't create islands of improvement in a sea of waste.

FIGURE 8.72
Kaizen improvements need to be linked to overall process improvement.

Kaizen Event Issues

Job Security Policy

Communicate that no layoffs will result from improvements in productivity directly related to a kaizen event. Kaizen focus is on eliminating waste, not people. People may be reassigned (unsatisfactory performers should not be protected by this policy; likewise, loss of business can result in reductions in all staffing levels).

Displaced Workers Policy

A list of specific new assignments should be prepared before the event, for example:

- Eliminate temporary employees (if any are currently employed)
- Grow business and enter into new markets (requiring people)
- Special projects, including facilitating future kaizen events and other continuous improvement efforts
- Cross-training

Pay Grade Differences Policy

When functional departments are combined into cells, operators may be required to perform functions that previously crossed pay grades. Some adjustment may be required—higher or lower.

Union Issues

Address and resolve any union issues before the event starts. These include overtime, work hours during the event, etc. Consider holding special meetings with union leaders at every stage of the process.

Team Leader Should Focus on Worker Participation

Team leaders should be encouraged to involve people from the shop floor throughout the event. The resulting gains are more likely to be sustained.

Plan for Advance Production

Be sure that you have adequate production to cover for the loss in production that may occur due to the absence of persons participating in the event.

Running a Successful Kaizen Event

A. Use "SPACER" to Begin Each Kaizen Event
Inform participants about
- Safety (restrooms, evacuation, housekeeping)
- Purpose
- Agenda
- Code of conduct (ground rules belong to the group… attendance, promptness, cell phones, etc.)
- Expectations (clarify purpose/obtain expectations)
- Roles, responsibilities, and rules (review sponsor, team leader, team members, facilitator roles, and rules slides)

B. Rules for Kaizen Events
- There is no rank among team members—one person, one vote
- Keep an open mind to change
- Maintain a positive attitude
- Never leave a silent disagreement
- Don't blame anyone for anything
- Respect one another
- There is no such thing as a dumb question
- Reject excuses, seek solutions
- Take action. Implement ideas immediately. "Don't let perfection get in the way of getting better." Just do it!

C. Characteristics of a Kaizen Event

A kaizen event is a highly focused, action-oriented, three- to five-day improvement workshop where a cross-functional improvement team takes immediate action to improve a specific process.

Team works (typically) three to five days full time

Resources Are Dedicated

Participants spend 100% of their time on the kaizen event. Participants should be treated as if they are on vacation from their regular responsibilities.

The Kaizen Event Is Well-Defined Beforehand

There is no time to define the purpose or scope in the event; therefore, business cases, goals, and charters must be well-defined ahead of time. Usually, the purpose of a kaizen event is to produce more with the same number of people or produce the same with fewer people.

Implementation Is Immediate

Implementation is completed as much as possible during the week of the event. Generating short-term wins greatly help justify the short-term costs involved and builds momentum. Don't let "perfect" get in the way of "better."

Management and/or Sponsor Participation Is Required

The area's top manager should kick off the event. The sponsor should "check in" at the end of each day to provide guidance and approval. At the conclusion of the kaizen event, management staff should be present so that the action items can be reported.

Following Up the Kaizen Event

The first thing to do is to evaluate the event. An example of an evaluation form is shown in Figure 8.73.

A kaizen event generally brings change. Change management aspects need to be carefully considered in following up the kaizen event.

Kaizen event evaluation form

Date: _____ Host plant: _____ Facilitator: _____
Name:_____ Event location:_____ Team leader:_____
(Optional) Product line: _____ Sponsor: _____

1. What was your overall impression of the event?

2. Describe the best or most useful part of the event:

3. What would you change about the event to make it more useful?

4. Would you like to participate in another event, yes or no, and tell why?

5. Did the event accomplish all that it could, or was there more that could have been done?

6. How were you treated? Could you give your opinions freely?

	Poor									Great
7. Please rate the facilitator.	1	2	3	4	5	6	7	8	9	10
8. Please rate the team leader.	1	2	3	4	5	6	7	8	9	10
9. Quality of training material:	1	2	3	4	5	6	7	8	9	10
10. Usefulness to the company:	1	2	3	4	5	6	7	8	9	10

Additional comments please:

FIGURE 8.73
Kaizen event evaluation form.

Change Management

Remember, change is exciting when it is done by us, threatening when it is done to us.

Commitment to Change

- 5% of people will embrace change—early adopters
- 5% will never change—concrete heads (anchor draggers)
- 90% will look for leadership—fence sitters

While traditional managers spend their efforts focusing on the anchor-draggers, they should be spending time with the early adapters, providing cover and support. The organization's focus should be on positive reinforcement, which promotes a forward shift among the fence-sitters.

Here is an example of the qouting process kaizen (as defined in days):

	Current State	Future State
Prior Quote	51	15
Quote Prior Quote (Can/Can)	63.75	5.75
Full SCM Quote	59	23.75
Customer Quote		< 5 Days

The above data focused in on the military qouting process; however, we have taken a straw poll and establish similar lead-times in the commercial qouting process. We are now off to the races with several immediate action items by the end of February

→ Implement daily pick-up of qoutes from web/fax and purchase fax cards for impacted employees.

→ Implement policy to extend quote validity from 90 days to align to spares price catalog

We are never standing still as we prep for several "mini-kaizen events" to take place in March/April.

Ray Grant and Kris Watson continue efforts with the future state map.

Customer-driven improvements

"This has been a powerful, productive and informative week. All of the participants had the synergy required to produce some very solid recommendations and suggestions that we are hopeful will be implemented

Having been here for a week, we are looking forward to a better partnering experience. Thank you, Eaton for opening yourselves to us."

Glenice Daniel-Chambers, DLA

What's next?

During the kaizen event, 4 processes were mapped sequentially—now to get started to arrive at the future state. We rated each process as to

▲ Achievability
▲ Impact to the customer
▲ Ease of implementation

The quoting process allowed for immediate benefits to the customer—so here we go...

Stan Allen, Glenice Daniel-Chambers (DLA), Elaine Juffs, and Time Thexton all work on the opportunities within the current for the qouting process

Who moved my cheese?
Maybe it's time to leave whatever stagecoach you are still on and enjoy another way of getting where you want to go—even if you have to go through the maze to get there!

FIGURE 8.74
Kaizen newsletter.

Why Do People Resist Change?

- Threat of uncertainty
- Fear of unknown
- Less security
- Less control
- Less power
- Less recognition
- Loss of identity
- Loss of contact
- More work
- More pressure

After the Kaizen Event—Sustain Improvements

After the kaizen event, the team should meet on a weekly basis to do the following:

- Review status of open action items (future kaizens)
- Review process metrics to ensure improvements are being achieved
- Discuss additional opportunities for improvements
- Continue to improve the process (PDCA)

Management should do 30-day reviews of the storyboard (idea board, Lean status board) to evaluate metrics, open items, and resolve any roadblocks to implementation. They should also provide recognition to the team as they reach milestones in implementation.

An example of a kaizen newsletter (*Cheese Report*), published biweekly, is shown in Figure 8.74.

A3 REPORT

In continuous improvement culture terminology, the A3 report is known as Mr. A3 or A3 San in Japanese. It is an 11 in. × 17 in. plain A3 size paper (Figure 8.75). It is a unique method of storytelling about Lean deployment projects. During the A3 process, the coach asks the learner to make an A3 report as follows:

Theme – title	A3 report layout
Background info – problem identification Which basic strategy/goal creates the need to take action on the problem now? What problem or issue is being addressed? Summarize the "business case" that makes this a priority. How will solving this issue and moving the proposed metrics impact the strategy goal?	**The future state** What will the proposed solution look like? How will the correction affect the current data and facts?
The current state What are the facts/data that support the need to take action? What is the current situation/process?	**Action plan (kaizen events)** What unresolved issue and obstacles need to be overcome? What is the priority of each? Who will be responsible for the action items? When is the expected completion date for each?
	The measure of success What are 3–5 key metrics that will prove the future state solution implemented actually solved the issues? Status tracking: Metrics at the start – initial state Progress to date – current state Targeted goal – future state objective

FIGURE 8.75
Sample A3 report layout.

- Define the background, what are we trying to do?
  - Describe what is happening and why it is important. The unfavorable trends, etc.
- Current state or situation
- Set future state actionable goals and targets
- Analyze current processes to understand the root causes of the challenge fully (five whys, Ishikawa diagram)
- Propose countermeasures
  - Develop actions to move toward a target condition
- Complete actions
  - Execute countermeasures to determine the effect. What, who, when, where, and the cost
- Standardize. Follow-up plan
  - Create standard work, set up monitoring or controls to maintain new standard work and new current state

The main purpose of the A3 process is to generate:

- A thorough and robust style of thinking
- Communication that focuses on factual data and vital information
- Collaborative and objective problem solving

A3 thinking can also be applied to the following projects:

- Project management
- Supplier capabilities
- New product/process concepts
- Capturing new knowledge

CORRECTIVE/PREVENTIVE ACTION SYSTEM

To be successful on the Lean journey and to implement a continuous improvement system is to have a sound corrective action system. This process ensures that a corrective and/or preventive action is taken when problems occur.

Preventive action is a proactive action taken on potential problems that could occur in the future. Therefore, a corrective action is necessary to drive out or to "switch off" the root causes and implement corrections to these issues.

Root Cause Analysis (RCA)

Every defect is a treasure, if the company can uncover its cause and work to prevent it across the corporation.

Kichiro Toyoda
Founder of Toyota

Sources from where the problems in an organization can be discovered:

- Noncompliances from internal and external audits
- Customer complaints, customer returns, field failures, and warranty cases

- Unfavorable trends in key performance indicators
- Failure to achieve a stated goal or objective
- Internal failures such as scrap, rework, or reprocessing
- Material nonconformances at the receiving (incoming inspection), in-process inspection, and final inspection
- Poor supplier performance, quality, and on-time delivery
- Root cause analysis and eight discipline processes

9

Continuous Improvement Process

Earlier, we have studied that the first requirement for improvements is the observation of the system or the business process. This knowledge will enable us to determine interrelated and interdependent processes. The next stage will be to change the way the work flows through these processes and make them least wasteful. Again, the process is a sequence of linked activities or tasks whose purpose is to produce a product or service for the customer.

Thus, if we want to make the process efficient and effective, we need to ensure that the process sequence runs or flows smoothly without obstruction of the non-value-added activities. We can describe this state as the values for the customer flowing smoothly. Such assurance is attained through the study of the work flow, called "work flow analysis" or value stream mapping.

VALUE STREAM MAPPING

We can evaluate the work flow or the value stream in two ways: flow charts and value stream mapping. A suppliers, inputs, process, outputs, and customers (SIPOC) chart is a form of a value-added flow chart, whereas a value stream map (VSM) is a flow chart on steroids—a flow chart blown up with details for each step of the process.

How to draw current state VSM (Figures 9.1 through 9.5):

Tips:
- Do a quick walk-through first to get a sense of the flow starting from incoming material to the final shipping (draw a simple block diagram).

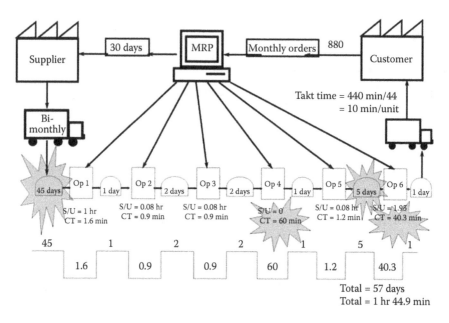

FIGURE 9.1

Elements of value stream map.

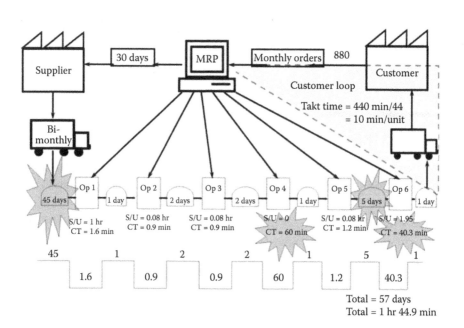

FIGURE 9.2

VSM customer loop.

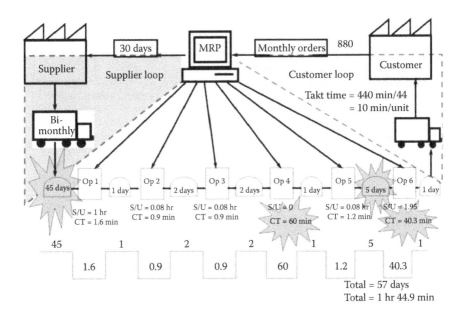

FIGURE 9.3
VSM supplier loop.

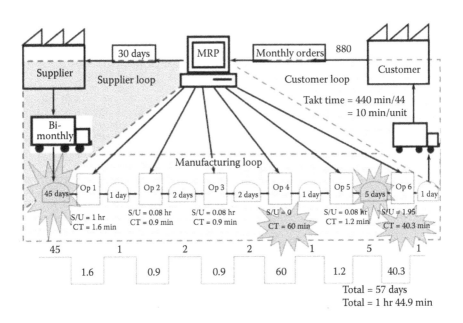

FIGURE 9.4
Manufacturing loop.

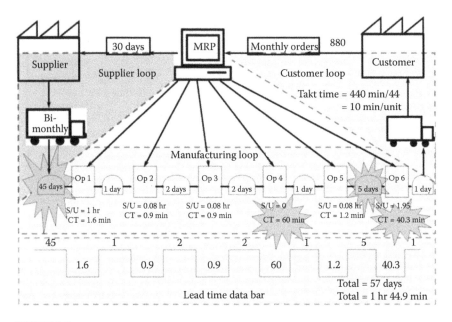

FIGURE 9.5
Lead time data bar.

- Begin the current state map nearest the customer, the shipping end, and work backward or upstream through the process.
- The information gathered for each step during the current stage of the map is as follows:
  - Persons per shift for each stage
  - Changeover time
  - Number of changeovers
  - Total cycle time
  - Value-added time versus non-value-added time
  - Lead time
  - Work in process inventory
  - Quality data such as scrap/defect rate

Customer Loop, Supplier Loop, Manufacturing Loop, and Lead Time Bar

Delay time in between operational stages is on the top line. This time determines total lead time. Operation cycle time (including a proportional setup time) is on the bottom line. Starbursts show the times where immediate corrective actions are needed (Figure 9.6).

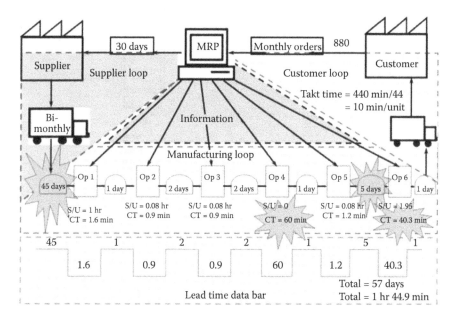

FIGURE 9.6
Information (MRP) loop.

Customer Loop, Supplier Loop, Manufacturing Loop, Lead Time Bar, and Information, Material Resource Planning (MRP) Loop

Exercise

Draw the current state map

- Divide the process into sections and assign team members to each section or value stream. Take a single product/service case as an example to start.
- Refer back to the business case goal statement to make sure you will be collecting enough data/process detail to help quantify to reasoning for improvements as stated.
- Share the information with the entire team to draw one current state map.
- Select a team member(s) to present the map to the entire shop/ department group.
- Select special (obvious) stages of the VSM with starburst icons so that these stages can be selected immediately for continuous improvement with respect to the delay and cycle times (Figures 9.7 and 9.8).

Starburst symbol

Mapping icons

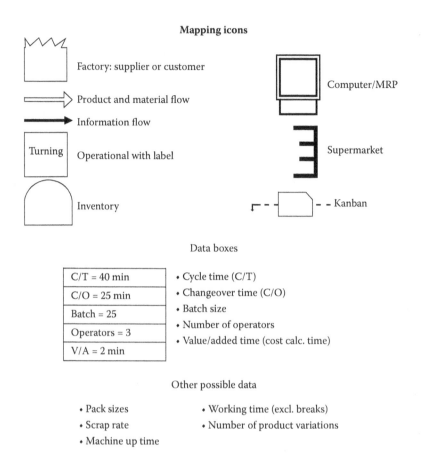

FIGURE 9.7
Value stream mapping icons and data boxes.

FUTURE STATE MAPPING

The following is an example of how the current state map enabled a Lean team to implement a change in the future state:

- A process control audit at the supplier facility reduced supplier delivery time from 45 to 30 days.
- Simple movement of plugging the responsibility after painting reduced the time of operation no. 6 from 60 to 20 minutes.
- Setup reduction and the addition of one honing machine reduced a delay of five days to three hours.

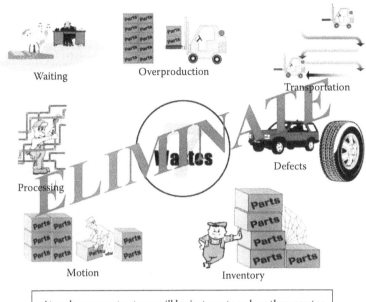

FIGURE 9.8
Manufacturing wastes.

- Operation no. 6 cycle time reduced to 10 minutes by moving part of the operation outside the process.
- This future state map will now become a current state map, and further continual improvements will be made on this chart. Thus, a plan–do–check–act (PDCA) cycle goes on.

So far, we have learned about the following Lean tools:

1. 5S
2. Total productive maintenance (TPM)
3. Standard work
4. Value stream mapping

Four more Lean tools follow:

5. Error proofing
6. Continuous flow or one-piece flow
7. Setup reduction
8. Pull system (Kanban system)

ERROR PROOFING

What is error proofing? It is the detection, containment, prevention, and elimination of a defective product at the source.

What are the steps?

- Eliminate potential defects which are capable of being but not yet in existence.
- Prevent defects from occurring.
- Detect the defect.
- Loss control.

What are the benefits?

- Reduced scrap, rework
- Reduce cost of quality
- More consistent output

What are the resource requirements?

A cross-functional team will take three days, including training. Error proofing is a systematic approach for anticipating and detecting potential defects and preventing them from reaching the customer (internal or external).

The following are the primary approaches for error proofing, in order of priority:

1. Eliminate potential defects: Design the potential for errors in the process.
2. Prevent defects from occurring: Find ways to make it difficult or impossible for employees to make an error in the first place.
3. Detect the defect: Even if an error occurs, it cannot be passed onto the next step of the process.
4. Loss control: Minimize the effect of the error.

Perform all the above-mentioned methods using the failure mode and effect analysis (FMEA).

Failure Mode and Effect Analysis

FMEA is the most systematic way to identify error proofing opportunities.

A little bit of trivia: FMEA used to be FEMA, a failure effect mode analysis, and it was developed by NASA to prevent catastrophic failures in space shots. During 1974, the U.S. Navy developed MIL-STD 1629 regarding the use of the potential failure mode and effect analysis (PFMEA). During the late 1970s, the automotive industry adopted it to control liability costs.

Failure Mode

Failure mode is the way in which a component or the process step fails "functionally" on a component/process level. Note: All FMEAs are considered potential FMEAs.

PFMEA is an approach

- to identify the ways in which a product or process could potentially fail
- to estimate the risk associated with causes
- to prioritize the actions to reduce the risk
- to evaluate the product design validation plan
- to improve the process in a preemptive manner (before failures occur)
- to prioritize resources to ensure process improvement efforts are beneficial to customer
- to document completion of projects
- to be a dynamic document for the current process control plan

Process PFMEA focuses on process inputs, and design PFMEA focuses on product function. The role of the PFMEA is to continually review, amend, and update the process for continual improvement.

PFMEA Inputs and Results

Inputs
- Process map
- Cause-and-effect matrix
- Process history
- Process technical procedures
- Others such as inputs from experts and lessons learned data

Results
- Prioritized list of actions to prevent causes and detect/minimize/eliminate failure modes
- History of actions taken (Table 9.1)

TABLE 9.1

FMEA Form (without Action Plan on RPN)

| | Identify Failures and Their Effects | | | Identify Causes, Controls, and Priorities | | | | | |
| | | | | | | Current Controls | | | |
Process Step Input	Potential Failure Mode	Potential Failure Effects	SEV	Potential Causes	OCC	Prevent	Detect	DET	RPN
What is the process step or input under investigation?	In what ways does the input go wrong?	What is the impact on outputs or external and internal customer requirements?	How severe effect is on customer?	What causes the input to go wrong?	How often does cause of failure mode occur?	What are existing procedures (inspection and test) to prevent/detect the FM?		How well can you detect causes of failure mode?	Risk priority number (SEV × OCC × DET)

Note: DET, detection; OCC, occurrence.

PFMEA—Step by Step

You will want to brainstorm the failure modes, effects, causes, and controls for each input with your team. There may be many failure modes for each step, so be sure to capture everything that is appropriate as you work through the process.

The rating scales are typically 1–10, with 1 being least severe, occurring least frequently and being totally detectable, and 10 being most severe, occurring most often and being totally undetectable. It is your team's choice, but a 1–10 scale is most common:

1. Layout the form with 17 columns as shown previously using an Excel spreadsheet.
2. For each process input, determine the ways in which the input can go wrong—the failure modes.
3. For each failure mode associated with the inputs, determine the effects of the failures on the customer—do not forget the internal customers.
4. Identify potential causes of each failure mode.
5. List the current controls for each cause or failure mode.
6. Create severity, occurrence, and detection rating scales.
7. Assign severity, occurrence, and detection ratings.
8. Calculate the risk priority numbers (RPNs) for each row and sort from high to low.
9. Determine recommended actions to reduce high RPNs.
10. Take appropriate actions and recalculate RPNs.

Consider that the PFMEA team consisting of the spray paint operator, the process engineer, the maintenance engineer, and the paint shop supervisor are analyzing the paint process where the first step is "paint." This will be entered in the first column of the FMEA form.

Step 1

Failure mode column: the way in which a specific process input fails—in this case the way in which paint operation can fail. If the failure mode is not detected, corrected, or removed, it will cause an effect to occur.

The failure mode can be associated with a defect or a process input variable that goes outside the specification (customer requirement).

For example, what can go wrong with the paint operation? It could be temperature too high (Table 9.2). Generally, it can be external customer focus, but it can also include downstream processes.

Step 2

Process potential failure effects: Brainstorm all the potential failure effects for each failure mode before moving on to the next step—the potential causes. By exhausting all the possible effects, you will move down the FMEA in a more orderly fashion, and you will not have to go back and add a lot of lines to the Excel spreadsheet where the FMEA is documented (Table 9.3).

Step 3

Cause: sources of process variation that causes the failure mode to occur. Again, brainstorm all the potential causes for one effect before moving on. You will save a lot of time in the Excel spreadsheet that way. As a tip, you may want to brainstorm the issues separately on Post-it Notes or a flipchart first before filling out the form. That way, you will not miss anything (Table 9.4).

Step 4

Current controls: systematic methods/devices in place to prevent causes or detect failure modes (Table 9.5). Which is more important to process improvement?

Prevention or Detection

We want to move to a culture that prevents rather than detects errors, so we want to develop controls that prevent rather than detect. PFMEA will focus on the important few.

We find the vital few by rating each row of failure mode/effect/cause/current control in terms of the following:

- Severity of the effect
- Occurrence of the cause
- Detection (capability of current controls)

TABLE 9.2

Step 1: Record All Process Potential Failure Modes for This Step

Process Step Input	Potential Failure Mode	Potential Failure Effects	SEV	Potential Causes	OCC	Current Controls Prevent	Current Controls Detect	DET	RPN
Paint	Temperature too high								

TABLE 9.3

Step 2: Record All Process Potential Failure Effects for This Step

| Process Step Input | Potential Failure Mode | Potential Failure Effects | SEV | Potential Causes | OCC | Current Controls | | | DET | RPN |
						Prevent	Detect			
Paint	Temperature too high	Paint cracks								

TABLE 9.4

Step 3: Brain Storm Causes for the Failure Mode to Occur

Process Step Input	Potential Failure Mode	Potential Failure Effects	SEV	Potential Causes	OCC	Current Controls Prevent	Current Controls Detect	DET	RPN
Paint	Temperature too high	Paint cracks		Broken thermocouple					

TABLE 9.5

Step 4: State Current Controls

Process Step Input	Potential Failure Mode	Potential Failure Effects	SEV	Potential Causes	OCC	Current Controls		DET	RPN
						Prevent (P)	Detect (D)		
Paint	Temperature too high	Paint cracks		Broken thermocouple			Inspection, test (D)		

Definition of Terms

- Severity (of effect)—importance of effect on customer requirements—concerned with safety and other risks if failure occurs (1 = not severe, 10 = very severe)
- Occurrence (of cause)—frequency with which a given cause occurs and creates failure mode. It can sometimes refer to the frequency of a failure mode (1 = not likely, 10 = very likely)
- Detection (capability of current controls)—ability of current control scheme to detect or prevent
  - the causes before creating failure mode
  - the failure modes before causing effect (1 = likely to detect, 10 = not likely at all to detect)
- When you begin your process improvement, you may not know how severe, how often, or how well you can detect a problem. In these cases, you may need to go out and collect data, but do not get bogged down in your FMEA. You may be able to validate presumptions with people in the process or based on historical data. You can always go back and make revisions. Remember, these tools are iterative. Doing an FMEA should not be a month-long or even a week-long exercise. Use it and move on.

Rating Scale Example

- Severity: severity of the effect (Table 9.6)
- Occurrence: probability of failure (Table 9.7)
- Detection: likelihood of detection by design verification (DV) program (Table 9.8)

Risk Priority Number (RPN)

- After rating, we get an output of the PFMEA
  - RPN
  - It is calculated as the product of three quantitative ratings, each one related to the effects, causes, and controls:

$$RPN = severity \times occurrence \times detection$$

TABLE 9.6

Severity Rank Numbers

Ranking	Description	General Criteria
1	None	No discernible effect
2	Very minor	Fit and finish does not conform; noticed by a discerning customer
3	Minor	Fit and finish does not conform; noticed by 50% discerning customers
4	Very low	Fit and finish does not conform; noticed by >75% discerning customers
5	Low	Operates at a reduced level of performance; customer is somewhat dissatisfied
6	Moderate	Operates at a reduced level of performance; some minor features inoperable; customer is dissatisfied
7	High	Operates at a reduced level of performance; customer is very dissatisfied
8	Very high	Inoperable; loss of primary function; customer is frustrated
9	Hazardous with warning	Affects safe operation and violates government norms with warning
10	Very severe/ fatal	Affects safe operation, can hurt limb; does not comply to government norms and without warning

TABLE 9.7

Occurrence (Probability of Failure)

Ranking	Description	General Criteria
1	Remote	Failure is unlikely <1 in 100,000
2	Low	Relatively few failures <1 in 10,000
3	Low	Relatively few failures <1 in 2000
4	Moderate	Occasional failures <1 in 1000
5	Moderate	Occasional failures <1 in 500
6	Moderate	Occasional failures <1 in 200
7	High	Occasional failures <1 in 100
8	High	Occasional failures <1 in 50
9	Very high	Occasional failures <1 in 20
10	Very high	Occasional failures <1 in 10

TABLE 9.8

Detection (Likelihood of Detection by DV Program)

Rank	Description	General Criteria
1	Almost certain	Design control will almost certainly detect a failure mode
2	Very high	Very high chance that design control will detect a failure mode
3	High	High chance that design control will detect a failure mode
4	Moderately high	Moderately high chance that design control will detect a failure mode
5	Moderate	Moderate chance that design control will detect a failure mode
6	Low	Low chance that design control will detect a failure mode
7	Very low	Very low chance that design control will detect a failure mode
8	Remote	Remote chance that design control will detect a failure mode
9	Very remote	Very remote chance that design control will detect a failure mode
10	Absolute uncertainty	No design control; cannot detect failure mode

Dynamics of the RPN

The team defines the rating scale (1–10) for the severity, occurrence, and detection ratings. You and your team choose the levels and numbers (as long as they are on a scale of 1–10).

How severe is it?
Not severe = 1
Somewhat = 3
Moderately = 5
Very severe = 10 (very bad)
How often does it occur?
Never/rarely = 1
Sometimes = 3
Half the time = 5
Always = 10 (very bad)
How well can you detect it?
Always = 1
Sometimes = 3
Half the time = 5
Never = 10 (very bad)

TABLE 9.9

Actions to Be Taken to Reduce RPN Continually

			Determine Actions				
RPN	**Actions Recommended**	**Responsible**	**Actions Taken**	**SEV**	**OCC**	**DET**	**RPN**
	What are the actions taken to reduce RPN?	Who is responsible for the recommended action?	What are the completed actions taken with the recalculated RPN?				

Using Excel, sort RPN numbers, keeping the highest number first.

Select all process steps where RPNs are greater than 100. (This is just an example. Actually, select the top 5 or 7 RPNs.)

Determine actions, responsible person, action completion resources, and time to reduce the RPNs.

Repeat actions by sorting the RPN and by selecting the top 5 RPNs and continue this PDCA cycle for continual improvement of the process or product (Table 9.9).

CONTINUOUS FLOW OR ONE-PIECE FLOW

What is it?

Continuous flow is defined as the movement of material from value-added process to value-added process without transport time or storage in buffers.

Processes are organized such that one person can build the entire product. If volume increases, additional people are added to match the takt time.

What are the steps?

1. Determine actual work elements
2. Design layout for flow
3. Distribute work content for flow
4. Run pilot and kaizen

What are the benefits?

- Reduced throughput time and reduced inventories
- More consistent output
- Making to order not to forecast
- Helps manufacturers achieve true just-in-time manufacturing

What are the resource requirements?

A cross-functional team from the area is needed for approximately 3 days for the plan and kaizen event. It should be preceded by VSM, 5S, and standardized work and may be preceded by TPM, setup reduction, and error proofing.

Let us find out if it is a batch or a single-piece flow.

Order Fulfillment Process

At a warehouse, it was mandatory to complete one order at a time, instead of several orders (batch) at a time. The associates had a difficult time understanding this concept until it was shown to them that the batch approach leads to "switcheroos"—that is, mistakenly putting an item in the wrong order. The one-piece flow approach avoided this defect altogether and led to shorter cycle times, enabling the order to meet shipping deadlines.

As shown in Figure 9.9, four operators take 20 minutes to process a batch of five parts, whereas the one-piece flow completes the job in eight minutes.

Continuous flow gives flexibility as follows: Below total work content = five minutes/part (Figure 9.10).

SETUP REDUCTION

What is it?

Setup reduction reduces the downtime due to changeovers so that batches can move quickly through the production process, making the company more responsive to customer needs and supply the customer what they want and when they want it.

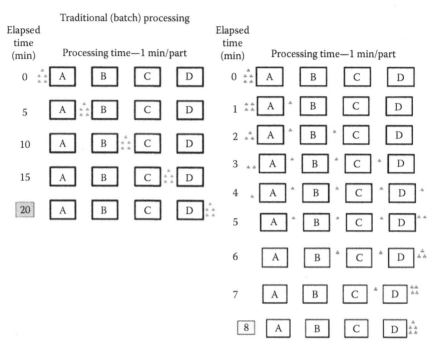

FIGURE 9.9

Continuous (one-piece) flow.

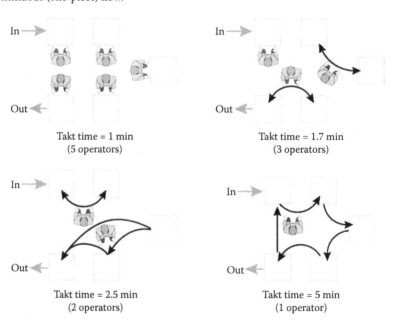

FIGURE 9.10

Continuous flow promotes flexibility.

What are the steps?

- Analyze the current setup process (video tape)
- Separate external and internal activities
- Convert as many internal activities as possible to external activities
- Eliminate/simplify the internal activities
- Eliminate/simplify the external activities

What are the benefits?

- Improvement of flexibility and productivity and more consistent output
- Reduction of batch sizes to improve delivery and reduce inventories

What are the resource requirements?

Cross-functional team, operators, supervision, engineering, and maintenance are needed for three days. A machine should be available to perform the changeover and to modify for improvements (Table 9.10).

PULL SYSTEM

Generally, the supply chains of the organizations work as "push" systems where products are produced based on a sales forecast and then stored in anticipation of the demand.

Lean production uses the "pull" rather than the "push" system to produce and ship the products. The pull system starts with the last operation

TABLE 9.10

Setup Reduction Phases

Period	Efforts/Time/Expense	Setup Time Reduction	Potential Activities
Immediate	Implement immediately; no expense	Up to 90%	Convert internal to external, organize tool storage, ensure tool fitness for use, organize hand tools and accessories
Short term	Relatively low/weeks or months/some expense	Up to 95%	Simplify location and clamping, eliminate adjustments, simplify approval, standardize procedures
Medium term	Relatively high/months or years/relatively high expense	Up to 99%	Design for reduced setup times: develop and implement standards, power clamps, powered handling, etc.

station and works backward through the system. Each operation requests the exact number of products needed from the previous operation. If the products are not requested, they are not produced and excess inventory is not generated.

Earlier, we have used an analogy to describe inventory to that of a river where rocks are covered with water and they cannot be seen. However, as the water (inventory) recedes, the rocks are exposed. Reduced inventory level shows the problems, and they can then be solved.

KANBAN

As we mentioned previously, for the pull system to work, communication between the work stations needs to be set up. This is made possible by the use of a signaling system called *Kanban*, a Japanese word meaning "signal" or "card." A Kanban card contains information such as part/product number, name, and quantity to be produced. It is attached to a container. When operators need a product from the operator or station before them, they pass on an empty container with an attached Kanban and take a full container. The operator who received the empty container with a Kanban now has the authority to produce the amount of products mentioned on the Kanban and fill the empty container.

The guiding principle: The methodology that nothing is produced until there is a "demand."

What are the steps?

- Flow product where you can
- Calculate Kanban size (maintain flow)
- Use container or marks for container location
- Use supermarkets at batch operations or where product must accumulate
- Calculate Kanban size (maintain flow)
- Calculate containers in the system
- Review as improvements are made or demand changes

What are the benefits?

- Simplified scheduling method
- Reduced inventories and throughput times
- More consistent production output

What are the resource requirements?

- Team of supervision
- Production planning and operators for two days; develop Kanbans and recalculate
- Normally preceded by continuous flow manufacturing, VSM, 5S, standardized work, TPM, setup reduction, and error proofing

Four types of Kanbans

- Production Kanban
- Supermarket Kanban. Supermarket is a storage for high-use materials that turn faster than the standard material handling cycle and is located close to the line to facilitate retrieval by an operator.
- In-process Kanban
- Supplier Kanban (Figure 9.11)

Table 9.11 presents a guide for an organization to assign Lean tool champions to appropriate department/process heads.

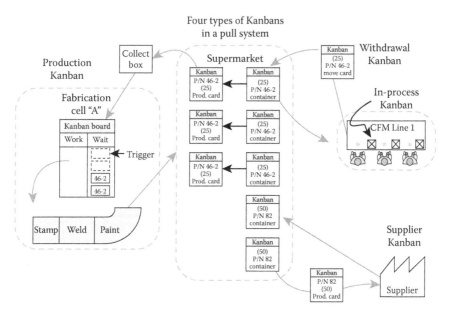

FIGURE 9.11
Four types of Kanban.

TABLE 9.11

Lean Tools and Their Suggested Champions

Lean Tool	Champions
• Pull (production Kanban)	• Production control manager
• Pull (supplier Kanban)	• Purchasing manager
• Error proofing	• Quality manager
• TPM	• Maintenance manager
• Setup reduction	• Manufacturing engineering manager
• Standard work and continuous flow	• Production manager
• 5S	• HR manager
• Value stream mapping	• Value stream mapping

Section IV

Lean Performance Measures and Performance Assessment

10

Lean Performance Measures

ON-TIME DELIVERY (OTD)

Definition

"OTD—zero days early, zero days late to the last date agreed to by the customer." It is a measure of an organization's ability to consistently meet customer expectations, a reflection of schedule stability and execution, a reflection of a stable management process, and a reflection of consistent materials, SCM, and HR management.

COST OF NONCONFORMANCE (CONC)

Definition

CONC is a "value of nonconformance cost" as measured by the inclusion of the "big seven" (numbers can be derived from a software such as Encore or compiled manually):

1. Product warranty
2. Rework
3. Scrap
4. Excess and obsolete inventory
5. Maintenance
6. Overtime (premium)
7. Premium inbound and outbound freight

Product Warranty

Product warranty is expressed with two subaccounts:

- Field failure costs
- As-received failure costs

Field Failure Costs (Product Warranty)

- Covers product deficiencies starting with the delivery of the equipment or service to the customer and excludes carrier damage.
- Specification conformance issues and field shortages clearly resulting from plant oversights or mistakes are to be included.
- Specification conformance issues and field shortages resulting from customer or field sales errors should be charged to a marketing budget or to a field sales error budget.
- Plant "product warranty" departmental costs associated with failure related activities should be included.
- Wages, benefits, and travel expenses for any plant or tech center salary or hourly employee assisting the warranty department should be captured and charged to a nonconformance account.
- Charges from internal engineering services or other subcontract services should be included.

As-Received Failure Costs

- Covers product nonconformances starting with the delivery of the products, equipment, or service to the customer, but excludes carrier damage.
- Includes dimensional, functional, cosmetic, and product marking issues as well as packaging, labeling, and packaging material contamination issues detected at receiving inspection or during use in the customer's facility.
- Wages, benefits, and travel expenses for any plant or tech center salary or hourly employee assisting in the resolution of the issue should be captured and charged to this account.
- Costs related to inspection, testing, or rework of the product by supplier, customer, or third-party resources are to be included. Transportation costs related to the inspection, testing, or rework are to be included.

- Penalty costs imposed by the customer, e.g., line stoppage, down-time, yard hold, etc., are to be included.
- Transportation costs to provide replacement product to the customer are to be included.

Rework

- Rework of returned product should include direct labor and material, including the costs for personnel to tear down and restock usable parts. Credit should be given to product warranty for the cost of parts reused and vendor expenses that are recovered.
- Labor and expense materials associated with internal production rework should be captured and assigned as nonconformance costs.

Scrap

- Scrap should be all deficient material that does not meet design specification and must be discarded. Blanked strip residues or punched slugs are "bill of material" items and should be excluded from the cost of nonconformance.
- Plants should maintain an effective system for capturing defective apparatus (DA). Full material, labor, and overhead should be included.
- Any recovery amount associated with scrapped material (i.e., copper, brass, aluminum, or steel) should not be credited against quality costs.

Excess and Obsolete Inventory

Excess inventory is the difference between actual inventory on hand and usage. It is calculated by a part number using three-year theoretical values derived from two-year actual values.

For example,

Three-year theoretical inventory: $10,000 (i.e., two-year actual inventory) × 3/2 = $15,000

Two-year actual usage = $8000

Three-year theoretical usage: $8000 (i.e., two-year actual usage) × 3/2 = $12,000

Excess inventory: three-year theoretical inventory—two-year theoretical inventory

Excess inventory: $15,000 – $12,000 = $3000

The company may allow to reserve 50% of the excess inventory cost; thus, the excess inventory reserve is $1500.

Obsolete inventory is the total of the following:

- Raw materials and work in process inventory that cannot be used to manufacture a readily saleable product, or for which there is no foreseeable use.
- Finished goods in a saleable condition that cannot be sold because there is no market demand.
- The reserve for obsolete inventory should equal the value of inventory on hand for three years that had no usage within the past year and is not being retained as a result of a contractual agreement with one or more customers. Actual obsolete inventory should be written off within the period in which it is identified.
- The value of excess and obsolete inventory that is written off plus any positive or negative change required in the reserve should be recorded as a cost of nonconformance.

Maintenance

- To start with, an organization must have a system in place to identify and exclude "preventive" maintenance costs and investments from their general maintenance accounts.
- For example, routine sharpening or rebuilding of tools to prevent defective production parts is preventive. Doing the same maintenance after manufactured parts are scrapped is considered operating in a failure mode.
- Maintenance employees that perform preventive maintenance on equipment should not be in general maintenance budgets.
- Any "general" maintenance agreements, such as routine machine and building maintenance, are to be considered preventive.

Overtime (Premium)

- The premium portion of overtime is considered a cost of nonconformance.

- This applies to production overtime only. Any overtime by engineering technicians associated with product development should not be included.
- If procedures and systems do not exist to allow the posting of the premium portion of overtime, then the site shall report 33% of the general overtime accounts as CONC overtime.

Premium Inbound and Outbound Freight

- Premium freight costs associated with moving material into or out of any facility to minimize the impact of failure or delay are to be recorded as nonconformance costs.
- Premium freight is defined as freight costs associated with air and ground expediting cost less the amount that is charged to customers.
- Expedited inbound and outbound premium freight used to resolve customer warranty errors or delays should be reported in this category.

The rest of the big seven costs are explained as follows:

1. Days on hand (DOH) = cost of goods sold/sales
2. Days sales outstanding (DSO) = outstanding receivables/sales
3. Days payable outstanding (DPO) = measures payments per agreed upon "terms"

These metrics (items 1–3) help us measure velocity of materials, conversion of material to sales, execution of contractual obligations, receivables aging, and payables aging.

They are a reflection of agility, a reflection of contractual conformance, and a quick assessment of cash assets and liabilities.

4. Cash flow return on capital (CFROC)
   - Measures our ability to generate cash utilizing our fixed capital assets
   - A reflection of our efficiency in utilizing our assets
   - A reflection of stability in scheduling key assets
   - A reflection of run rate and equipment utilization
   - A reflection of a systematic and right-sized appropriation
   - A reflection of a systematic approach in determining footprint or plant layout

5. Productivity

Productivity can be defined as cost per units produced. In manufacturing industries, it is easy to define productivity because production numbers are available and the expenses can be determined. In the case of $100,000 produced over 1 month with a combined total man hours of 5000, the calculation would be as follows: $100,000/5000 = $20 per hour productivity.

- Measures our ability to achieve production standards
- A reflection of our consistent and well-aligned planning
- A reflection of stability in scheduling key assets
- A reflection of run rate and equipment utilization
- A reflection of a thorough manufacturing process understanding
- A reflection of a systematic approach in determining footprint

6. Cycle time

For details on cycle time, see Chapter 8 on standard work.

7. Customer returns (DPPM) and supplier rejects (DPPM)

This is the ratio of total plant-verified customer returns reported in the month to total units sold during the month expressed in "parts per million."

Formula: (Number of verified units rejected at Customer facility/Number of units shipped) × 1,000,000.

Note: Units that are analyzed and found to be not the responsibility of the company are excluded from this calculation.

11

Lean Tool System Assessment

VALUE STREAM MAPPING

Value stream mapping (VSM) is a visual tool that diagrams the product, material, and information flows from the customers' order to the receipt of finished goods. The basic requirements are a current state map, a future state map, an implementation plan, and metrics to measure and communicate (posted) your progress.

Value	Conditions
1 Point	Value stream map is being developed.
2 Points	All product families have been formally identified.
	At least one current state and future state value stream maps are developed.
3 Points	Multiple current state and future state value stream maps are developed.
	Implementation plans are documented and in progress.
	Metrics to measure the progress have been selected and are being used.
4 Points	At least one future state implementation plan is completed.
	All product families have current state and future state value stream maps developed.
	All product families have documented implementation plans.
5 Points	All product families have one completed future state value stream map.
	New future state maps and implementation plans have been developed for all product families.
	A systematic process is in place to improve the current system to identify product lines/families and mapping the current and future states.

Score	0
Comments	

5S

5S is the basis for Lean and the foundation for a disciplined approach to the workplace: (1) sort, (2) straighten, (3) shine, (4) standardize, and (5) sustain. This step-by-step approach helps clean up the place of unnecessary items, dirt, and clutter and organize everything based on the principle of "place for everything and everything in its place." 5S implementation benefits office areas as much as the manufacturing shop floor. 5S can be implemented in a small area around a machine, in an entire department, and throughout the plant. While assessing a facility (plant or a department), look for extent of deployment.

Value	Conditions
1 Point	5S program has been started.
2 Points	Sort and straighten have been completed in 50% of the areas of the facility, and evidence is available to support it.
	Criteria for disposal of not-needed items have been developed, and items have been identified and tagged.
	Designated area for holding not needed items is established, and all these items are removed to this area. This area can be temporary.
	Location for every needed item has been established and labeled.
	Outline for equipment, supplies, common areas, and safety zones are marked.
	Shadow boards are deployed, where appropriate.
	Visual display boards are in use.
3 Points	Sort, straighten, and shine have been completed in the entire facility, and evidence is available to support it.
	Major causes of contamination that can make it difficult to keep the area clean are identified and removed.
	Documented cleaning inspection procedures are implemented.
	Cleaning is a part of everyday activity.
	Visual display boards are maintained in an orderly and timely manner.
4 Points	Sort, straighten, shine, and standardize have been completed throughout the facility, and evidence is available to support it.
	Documented standard practices and routines are established for systematically repeating the first three Ss.
	Procedures and forms are created and implemented that help regularly audit the status of first three Ss.
5 Points	Sustain has been completed for the entire facility, and evidence is available to support it.

Management ensures that 5S activities are a habit for all and that the standards are met, through personal involvement and assessments.

5S standards are a part of daily work and are linked to the other relevant initiatives. A systematic process is in place to continuously evaluate and improve these standards.

Score	0
Comments	

STANDARDIZED WORK

Standardized work is the optimum combination of operators, machines, and materials to ensure that a task is completed the same way every time with minimum waste and at market pace or takt. In most cases, there is more than one way of performing a task, and one may be more efficient than the other. A team of appropriate people (operators, line leaders, and engineers) develops and documents standardized work—the most efficient way—after evaluating all the different ways of performing the same task. The objective of standardized work is to make sure operator productivity and equipment utilization are simultaneously optimized and also that the inventory is minimized. Using five different forms, this objective is achieved by establishing a relationship between takt and cycle time, balancing the work content of all the operators and equipment in a cell, and establishing appropriate quantities for the pull system.

Value	Conditions
1 Point	Standardized work is currently being developed, and some documentation is completed.
2 Points	Some operations are performed in a standardized manner, as a result of individual operator efforts.
	Standard work layouts are implemented in some cells.
	Some standards are documented and followed by all the operators.
	Some cycle times are less than or equal to takt.
	Some work is balanced, but not to takt.
	Some production display boards are used to track hourly production status.
3 Points	Several cells and flow lines have implemented standardized work, where documented standardized work charts are available for all the operators.

Standardized work has been developed for most common takt.

Operator balance charts are implemented in some cells.

Several cells are beginning to use takt as the pacing element.

4 Points   Standardized work is implemented, and takt is followed by most of the cells and flow lines.

Opportunities have been identified for removing waste from the complete value stream, by standardizing the work that affect intercell activities and inventories.

Evidence exists that operators are beginning to generate standard work improvements. Takt is pacing element throughout the facility.

There is visible evidence that resource adjustments utilize standard work to balance changes in takt.

5 Points   Standardized work is implemented through all the cells and flow lines for all the possible takt.

The entire value stream is being continuously optimized (waste eliminated) by standardizing all the intercell activities and inventories.

Evidence exists that operators are generating and maintaining standardized work.

Standardized work has to be developed for all the possible takt.

A systematic process is in place to continuously evaluate and improve these standards.

Score	0
Comments	

TOTAL PRODUCTIVE MAINTENANCE (TPM)

Because Lean requires just-in-time supply of all the components and subassemblies, if the production equipment breaks down or fails to produce at the expected rate or quality, Lean product lines can suffer. Overall equipment effectiveness (OEE) is used to measure the equipment's performance in this regard. To make sure that the production equipment performs at or better than the expected level, they should undergo TPM. TPM begins with updating the equipment's condition to as good or better than the new machine, and it develops and implements procedures and checklists that ensure the maintenance of this level of the equipment's condition for a long period of time. A good TPM program also requires well-implemented 5S practices. World-class manufacturing plants consistently maintain their average OEE higher than 85%. Critical equipment in these plants is maintained at an OEE of 95% or higher. World class plants do not need to measure OEE all the time, on all the machines. The OEE of average manufacturing plants is usually between 35% and 50%.

Value	Condition
1 Point	Routine maintenance is carried out regularly, and preventive maintenance is practiced to some extent.
2 Points	A systematic preventive maintenance program is in place.
	OEE is being implemented, and baseline is being established.
	TPM implementation plan is being developed for equipment identified as critical on value stream maps.
3 Points	Initial cleaning and inspection completed on all the critical equipment.
	Most of the causes of contamination and inaccessible areas have been eliminated on all the critical equipment.
	Operator-performed cleaning and inspection standards are developed and deployed for all the critical equipment.
	Improving trends in OEE measure are evident.
4 Points	All the causes of contamination and inaccessible areas are eliminated on all the critical equipment.
	Training on equipment's functions, controls, and systems are completed for all the critical equipment.
	Regular audits of all the critical equipment are conducted for deployment of autonomous standards.
	All the critical equipment identified on the VSM maintain optimum OEE levels.
5 Points	Autonomous cleaning and inspection standards are deployed throughout the plant.
	Regular audits of all the production equipment are conducted for the implementation of autonomous standards.
	OEE is used for periodic monitoring of production equipment, and results are consistently sustained at optimum levels.
	The Six Sigma approach is being applied for predictive maintenance.
	A systematic process is in place to evaluate and improve OEE monitoring, TPM, and procedures deployed for cleaning and inspections.

Score	0
Comments	

ERROR PROOFING

Error proofing is a systematic approach to prevent potential defects from leaving the area in which they are produced. During error proofing, all the opportunities for defects (key characteristics) are identified, proactively, and the causes eliminated or 100% verification means are put in place either to prevent the error from occurring or to detect a defective product. This will lead to corrective actions taking place and getting closer to the "goal of zero defects."

Value	Conditions
1 Point	Inspection or testing is used to capture some of the potential defects.
	Root causes of defects are investigated infrequently.
2 Points	Systematic approach (such as design and process PFMEA) is being deployed to identify the potential defects affecting the external customers.
	Causes for many of the identified defects are either eliminated or detected and contained at the source.
3 Points	Systematic approach (such as design and process PFMEA) is deployed to identify all the potential defects affecting the external customers and many of the potential defects affecting the internal customers.
	All the causes of potential defects affecting the external customer have been eliminated or detected and contained at the source.
4 Points	Systematic approach (such as design and process PFMEA) is deployed to identify all the potential defects.
	Causes of all the potential defects have been eliminated or detected and contained at source.
5 Points	A systematic process of elimination, prevention, detection, and loss control, in that order of preference, is applied to all the potential errors.
	Error proofing is actively implemented during product and process development.
	A systematic process is in place to evaluate and improve the deployed error identification means and error proofing methods applied.

Score	0
Comments	

SETUP REDUCTION

Setup is the time during which equipment is being set up for the next order and not producing parts—as defined by the time between the last good piece of the previous setup and the first good piece of the current setup. Lean uses a small quantity of controlled inventory; this means smaller batch sizes and frequent runs. Thus, reducing the setup becomes very important for a product line to be Lean. Setup reduction is a systematic approach used to reduce the nonproductive time during setup.

Value	Condition
1 Point	Many factors affect the overall setup time because the setup processes have not been standardized and documented.
2 Points	Some setups are evaluated (e.g., analyzing video tapes), and related activities are standardized and documented and standards are being followed.
	There is a metric defined to measure setup time for critical equipment—as identified in the value stream map.
3 Points	Most setup-related activities are standardized and documented, and standards are being followed for critical and subordinate equipment.
	Setup activities are systematically evaluated (e.g., analyzing video tapes), and attempts are made to move as many internal activities (when machine is idle during setup) to the external setup process (when machine is producing good parts).
	Attempts are made to reduce the time taken by the remaining internal activities.
4 Points	All possible setup-related activities are standardized and documented, and most operators follow the standards—for critical and subordinate equipment.
	Setup activities are systematically evaluated, and all the potential internal activities are moved to the external setup process, resulting in lead time and inventory reduction and greater flexibility.
	All the remaining internal activities are being continuously optimized.
	Efforts are made to continuously optimize the external activities.
5 Points	All internal and external activities are being continuously optimized for critical and subordinate equipment.
	Setup-related standards are deployed throughout the plant and used by everybody, resulting in the ability to run all parts every day.
	Systematic continuous improvement process is implemented, which helps achieve progressive reduction in setup time.

Score	0
Comments	

CONTINUOUS FLOW

Continuous flow is defined as the movement of material from one value-added process to another without transport time or storage in buffers in a spirit of "make one—move one." In a continuous flow environment, the rate of production of the entire product stream exactly matches the customer demand (takt). Continuous flow environment has optimized equipment, manpower, and space through effectively linked manufacturing cells, balanced workload for all the workstations and the operators (for various takt). Also, continuous flow environment provides flexibility through U-shaped cell designs, cross-trained work force, and ability to flex. Continuous flow environment would have documented all the task details and are easily visible to the operators. Also visible is a display that indicates the customer demand rate and achieved production rate in real time.

Note: Need to define "cell" in the glossary to encompass the concept of process.

Value	Conditions
1 Point	Processes are separated by a considerable amount of work in process (WIP).
	Processes are separated by geographic distance that requires extensive and complex transport path for parts.
	Machines are grouped based on equipment type, not process flow.
2 Points	Some cells are formed to enhance operator efficiency, and it effectively reduces the parts transport or the controlled WIP.
	Evidence exists that a clear understanding of customer demand has been included into a cell design to promote flexibility.
	Evidence exists that cell designs are defined by value stream maps.
3 Points	Several cells exist that internally flow "one piece."
	Some cells are linked together to flow the product through the entire product value stream.
	Production rates of the cells are variable based on customer demand.
4 Points	Most production equipment is based on the value stream map in cells.
	Most cells are linked to achieve continuous flow, with minimum controlled WIP between them.
	Production rate matches the takt.
	Operators are cross-trained within the cell's activities, and the cell is visually displaying a skills matrix.
5 Points	All the possible equipment is in cells and linked to flow based on the value stream map.
	Each cell is capable of adjusting the number of operators based on the takt.
	Each operator is cross-trained to perform several activities and also flex up or down, as needed.
	Kanban-controlled WIP are used only where the process does not allow continuous flow (e.g., batch heat treatment process).
	A systematic process is in place to continuously evaluate and improve the standards.

Score	0
Comments	

PULL SYSTEM

Pull is a system of cascading production and delivery instructions from downstream to upstream activities in which nothing is produced by the upstream supplier until the downstream customer signals a need. A variety of simple visual methods (Kanban) are used by the customers to signal the upstream supplier to supply/produce just in time (JIT). A signal to indicate the customer (outside customer) order is given to only one internal process, which is called a pacemaker (ideally to the last process). Ideal pull system requires that the setup times are very low to produce small batch sizes, without economic penalty, and the quantity of those batch sizes or work in process (WIP) is a good indicator of the pull system's level of implementation. In a pull system environment, if no signal were received from the downstream customer, equipment and operator would be idle. In environments without the pull system, parts are produced as fast as one can produce and pushed to the downstream customer, regardless of need.

Value	Conditions
1 Point	Some evidence of a pull system exists.
	Efforts are being made to ensure that inventory is not stored in random amounts with many different component quantities being delivered at many different areas (typical batch warehouse).
	Most of the processes receive orders from a typical MRP/ERP-type system.
2 Points	Some production Kanban are used, but primarily for large batch sizes, to and from supermarket/stores. The maximum batch size is predetermined by the sales order quantity.
	Some processes get signals from the downstream process, to supply or start production.
3 Points	Products/parts requiring batch processing and buffer inventory (supermarkets) are produced and supplied using pull signals (Kanban).
	Some suppliers receive JIT pull signals or Kanban to signal the need.
	Constant attempts are made to reduce the batch sizes (internal and external customer).
	A methodology is in place to calculate appropriate Kanban sizes on an ongoing basis (setup reduction times are reviewed relative to lot sizes).

In-process Kanbans are used in many places, which ensure pull system between processes.

4 Points   The only parts being stocked in the supermarket(s) are those being produced/supplied in batches, due to limitations of the production equipment or supplier.

The quantity of parts in the supermarket have been systematically minimized.

Conveyance of parts is well planned, in terms of timing, sequencing, and quantities.

All "A-classified" material suppliers get orders through a type of JIT Kanban signal.

Most of the internal processes throughout the value stream initiate production/delivery based on pull signals.

Appropriate means (Andon panels, lights, sirens, etc.) to attract attention to out-of-parts condition are installed to avoid any line stoppages.

5 Points   Material movement throughout the plant value stream is based on pull signals linked to real demand.

Movement of material between processes is minimized, and WIP buffers are minimized or eliminated to the smallest practical delivery quantity but not more than one small container, box, or part, where applicable.

Customer orders are sent to the "last process" triggering the pull of components from the entire value stream.

A systematic process is in place to continuously evaluate and improve these standards.

Score	0
Comments	

TOOL ASSESSMENT SUMMARY RADAR CHART

A radar chart can be made by inserting the assessment score for each tool using Excel.

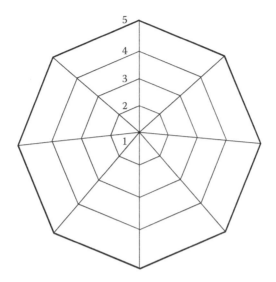

LEAN ASSESSMENT

Element	Score
Value stream mapping	1
5S	1
Standardized work	1
TPM	1
Error proofing	1
Setup reduction	1
Continuous flow	1
Pull system	1
Total	8
Average	1.0

Appendix 1: Attributes of a Truly Lean Organization

Finally, you know you are Lean if

- You can see what is going on without asking
- Everyone
  - shows up and
  - "pulls the cord"
- Supervisors are
  - checking the process, checking the process, checking the process, checking the process, checking the process
- Everyone in the organization is regularly on the shop floor
- Every process is standardized and documented
- Problems are embraced, not shunned
- You cannot find the WIP
- The janitor has standardized work and a production pacing board
- You can eat off the production machines
- No one can remember when a shipment date was missed
- People ask "Why? Why? Why? Why? Why?" and no one gets mad at them
- Management is concerned that they are gaining market share too quickly
- The no. 1 selling book at Crossword is about your company's success
- Companies from around the world are learning new words … in your language
- Customer buyers start ordering weird products with standard lead times … just to win bets with friends
- Your company moves production back into the original garage, but with $500 million annual sales
- The only place you can find any dust is in the scrap bins
- You think you find a problem. No kanbans. Then you learn everything runs in single piece flow
- The parking lot is one big FIFO lane

- You could not buy a Coke with your monthly cost of nonconformance (CONC)
- You have a drive-through lane to take your customer's order and then load it into the truck
- Machine manufacturers come to look at their TPM'ed machines in your plant before designing their next model

Appendix 2:
Job Description of a
Lean Supervisor

A. JOB DATA

Job position: Supervisor (SV)
Code:
Function: Supervisor
Area: Production

B. SUMMARY (DESCRIBE BRIEFLY THE POSITION GOALS)

Coordinate the area that is responsible, managing and giving orientation for operators under his responsibility, to develop activities according to plant goals.

C. MAIN RESPONSIBILITY (DESCRIBE ACTIVITIES DONE CONSIDERING FOUR BASIC QUESTIONS: WHAT DOES HE DO? HOW DOES HE EXECUTE? WHICH KIND OF EQUIPMENT DOES HE USE? WHY DOES HE DO IT?)

Participate in the manager meeting and define strategy for his area to achieve the goals defined by the plant manager.

Reinforce out-of-line resources provision (man, maintenance, material, safety, and environment) to reduce the line problem impact.

Participate to develop new processes (equipment, tools, materials, and methods), and be responsible to introduce it in line.

Consolidate production results, and prepare and present it in the manager meeting.

Daily analyze the production results and take action when there is a need to recover delays (quantity, overtime, HR issues, etc.).

Analyze production, quality, safety, cost, and delivery issues, and take actions according to TPS to improve the productivity and operators environment.

Daily check the quality and the productivity on each line, checking problems and making follow up in the countermeasures.

Check if group leaders and team leaders are orienting operators according to the operation standard, standardized work, using check sheets correctly, and 5S.

Control the cost in his area.

Give orientation about MESH (environment, safety, and health), TPS, and the Toyota way.

Participate and encourage your staff to develop Kaizen Circle activities.

D. QUALIFICATION

D1—Obligatory qualification
  Academic background: completed college/university
  Experience: more than five years in the industrial operation
  Language: Portuguese and English (basic)
D2—Desired qualification
  Academic background: production or mechanical engineer, high school completed at technical school (SENAI)
  Experience: group leader experience
  Language: Japanese
  Additional knowledge: proactive, organized, and good relationship among operators and areas
D3—Training and skills
  According to the manufacturing area

E. ORGANIZATIONAL STRUCTURE

Plant manager	Second-level manager
Department chief	First-level manager
Supervisor	Job description title
Other staff below this manager	Direct staff below
–	–
Plant manager	Human resources
__/__/____	__/__/____

Appendix 3:
Job Description of a Lean
Team Leader

A. JOB DATA

Job position: Team leader (TL)
Code:
Function: Team leader
Area: Production

B. SUMMARY (DESCRIBE BRIEFLY THE POSITION GOALS)

Coordinate the job in the area that is responsible, managing, and giving orientation for operators under his responsibility to develop activities according to plant goals.

C. MAIN RESPONSIBILITY (DESCRIBE ACTIVITIES DONE CONSIDERING FOUR BASIC QUESTIONS: WHAT DOES HE DO? HOW DOES HE EXECUTE? WHICH KIND OF EQUIPMENT DOES HE USE? WHY DOES HE DO IT?)

- Distribute the activities according to the daily plan, and advise operators about procedures and technical issues to perform according to the standard.
- Provide equipment, tools, materials, and accessories to run the area.
- Give orientation for operators about the machine stop, to use their time in the best way.

- Update the information about production plans and communicate it for every operator, informing production changes (quantity target, overtime, HR issues, etc.).
- Collect production data to group leader (manufactured quantity by line, machine problems, productivity, etc.).
- Read and understand product and manufacturing process specifications to instruct new operators or when they have any doubt.
- Prepare and deploy operation standards, standardize work, check sheets, and define 5S standards.
- Check if operators are filling correctly the check sheets.
- Give orientation about MESH (environment, safety, and health), TPS, and the Toyota way.
- Participate and encourage your staff to develop Kaizen Circle activities.

D. QUALIFICATION

D1—Obligatory qualification
- Academic background: completed high school
- Experience: more than two years in the industrial operation
- Language: only Portuguese

D2—Desired qualification
- Academic background: high school completed at technical school (SENAI)
- Experience: more than three years of work in the manufacturing activity
- Language: English, basic level
- Additional knowledge: machine NC control program, computer use, forklift and crane certification

D3—Training and skills
- According to the manufacturing area

E. ORGANIZATIONAL STRUCTURE

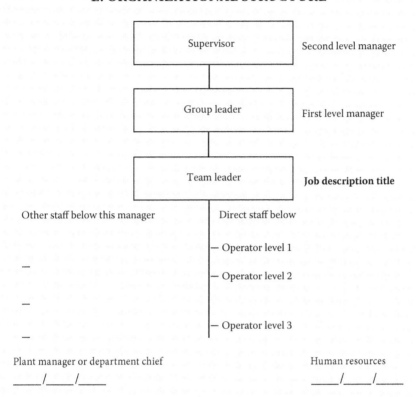

Supervisor	Second level manager
Group leader	First level manager
Team leader	**Job description title**

Other staff below this manager | Direct staff below

— Operator level 1

—

— Operator level 2

—

— Operator level 3

—

Plant manager or department chief

_____/_____/_____

Human resources

_____/_____/_____

Conclusion

During the last quarter of the 20th century, nearly everyone in senior management thought that manufacturing operations had been perfected. Henry Ford's great innovation, the moving assembly line, had been refined over the five decades after 1910, had served as the magic bullet during World War II, and by the mid-1960s was operating efficiently, at mass production scale, in a wide range of industries around the world.

Silently, in Japan, Taiichi Ohno and his engineering colleagues at Toyota were perfecting what they named the Toyota Production System, which we now know as Lean production.

Initially, Lean was best known in the West by its tools: for example, kaizen workshops, where frontline workers solve knotty problems; kanban, the scheduling system for just-in-time production; and the andon cord, which, when pulled by any worker, causes a production line to stop.

In more recent years, this early understanding of Lean has evolved into a richer appreciation of the power of its underlying management disciplines: putting customers first by truly understanding what they need and then delivering it efficiently, enabling workers to contribute to their fullest potential, constantly searching for better ways of working, and giving meaning to work by connecting a company's strategy and goals in a clear, coherent way across the organization.

Similarly, these companies are discovering that Lean can supply powerful insights about the next frontiers of energy efficiency. Toyota itself is pushing the boundaries of Lean, rethinking the art of the possible in production-line changeovers. For example, bring customer input more directly into factories. Leading service-based company such as Amazon.com is extending the value of Lean further still, into areas beyond manufacturing, such as fulfillment centers.

RETAIL BANKING

Lean has transformed how leading companies think about operations—starting in assembly plants and moving more recently into services such as retail banking because it involves a physical process, such as an assembly line, the handling of paper checks, and credit card slips, which lends itself readily to Lean manufacturing techniques. Moreover, their impact can be dramatic: the faster a bank moves checks through its system, the sooner it can collect its funds and the better its returns on invested capital.

HOSPITALS

It is agreed that a hospital is not an automobile factory, and people (the patients and the hospital staff) are less predictable than car parts. Nevertheless, hospitals can often reduce a large amount of their variability in operations, such as prescription distribution, use of consumables, and patient registration and outpatient processes.

AIRLINES

Aircrafts worth millions or more routinely sit idle at gates. Turnaround times between flights typically vary by upward of 30%. Lean techniques cut hours to minutes with an improved changeover system.

RESTAURANTS

Lean techniques seek to improve product and service quality while simultaneously reducing waste and labor costs. For food service operators, the additional trick is to link such improvements to customer loyalty.

DESIGNING PRODUCTS FOR VALUE

Data are now available through machine telemetric devices having embedded sensors. These small data sensors monitor installed equipment in the field and give companies insight into how and where products are used, how they perform, the conditions they experience, and how and why they break down. The next step is to link this information back to product design and marketing—for example, by tailoring variations in products to the precise environmental conditions in which customers use them.

Savvy companies will use the data to show customers evidence of unmet needs they may not even be aware of and to eliminate product or service capabilities that are not useful to them. Applying Lean techniques to all these new insights arising at the interface of marketing, product development, and operations should enable companies to make new strides in delighting their customers and boosting productivity.

Glossary

5S: stands for five Japanese words that begin with the letter "S." The words as translated in English are as follows: Seiri = sort, Seiton = straighten or set in order, Seso = shine, Seiketsu = standardize, and Shitsuke = sustain. All together, they mean orderly, well-organized, well-inspected, clean, and efficient workplaces.

5-Whys: a simple process of determining the root cause of a problem by asking "why" after each situation to drive deeper in more detail to arrive at the root cause of an issue.

7 Wastes: originally identified by Taiichi Ohno. These are (1) overproduction, (2) waiting, (3) transportation, (4) overprocessing, (5) stock on hand, (6) movement, and (7) making defective product.

8D: a popular method for problem solving because it is reasonably easy to teach and effective. The 8D steps and tools used are as follows: D0, prepare for the 8D; D1, form a team; D2, describe the problem; D3, interim containment action; D4, root cause analysis (RCA) and escape point; D5, permanent corrective action; D6, implement and validate; D7, prevention; and D8, closure and team celebration. This process is known as Global 8D by Ford.

A3: a report prepared on an $11'' \times 17''$ plain paper by the owner of the issue. The PDCA format is used. It gathers current information and analysis, creates goals and metrics, and builds buy-in from stakeholders.

Andon: a Japanese word meaning light or lantern. It is a form of communication for abnormal conditions or machine malfunction. It often resembles a stop traffic light where red = stop, yellow = caution, and green = go. Another form can be an andon cord, which is pulled by the operator to communicate an abnormal situation.

AS 9100: a widely adopted and standardized quality management system for the aerospace industry.

BE: business excellence.

Black belt: a professional who can explain and practice Six Sigma philosophies and principles, including supporting systems and tools.

Budget: an estimate of costs, revenues, and resources over a specified period, reflecting a reading of future financial conditions and goals.

Business process reengineering (BPR): the analysis and redesign of workflow within and between enterprises. BPR reached its maximum popularity in the early 1990s.

CA: corrective action taken to eliminate the cause of the nonconformity.

Cause and effect diagram: this diagram-based technique helps us identify all of the likely causes of the problems faced in working environments.

Changeover: setting up a machine or production line to make a different part number or product.

Changeover time: the time from the last good piece of the current production run to the first good piece of the next run.

Constraint: anything that limits a system from achieving higher performance. It is also called a bottleneck.

Continual improvement: continual indicates duration of improvement that continues over a long period, but with intervals of interruption, for example, the plant modification disrupted by logistics/traffic for nearly two years.

Continuous improvement: an approach of making frequent and small changes to a process whose cumulative results lead to higher levels of quality, cost, and efficiency.

Countermeasure: corrective action taken to address problems or abnormalities.

Customer: a party that receives or consumes products (goods or services) and has the ability to choose between different products.

Cycle: a sequence of operations repeated regularly.

Cycle time: the time for one sequence of operations to occur.

Effectiveness: the degree to which objectives are achieved and the extent to which targeted problems are solved. In contrast to efficiency, effectiveness is determined without reference to costs and, whereas efficiency means "doing the thing right," effectiveness means "doing the right thing."

EFQM: European Foundation for Quality Management.

Equipment availability: the percentage of time equipment (or process) is available to run. This is sometimes called "uptime."

Error proofing: see Poka Yoke.

External setup: procedures that can be performed while a machine is running.

FAI: first article inspection.

Failure mode and effects analysis (FMEA): a step-by-step approach for identifying all possible failures in a design, a manufacturing or assembly process, or a product or service. "Failure modes" means the ways, or modes, in which something might fail. Failures are any errors or defects, especially ones that affect the customer, and can be potential or actual. "Effects analysis" refers to studying the consequences of those failures.

FIFO: "first-in, first-out"; in other words, material produced by one process is consumed in the same order (FIFO) by the next process.

First pass yield (FPY): defined as the number of units coming out of a process divided by the number of units going into that process over a specified period. Only good units with no rework are counted as coming out of an individual process. Also known as throughput yield (TPY).

Fishbone diagram: the fishbone diagram identifies many possible causes for an effect or problem. It can be used to structure a brainstorming session. It immediately sorts ideas into useful categories. The major categories of causes of the problem are methods, machines (equipment), people (manpower), materials, measurement, and environment.

Flow: the completion of steps within a value stream so that product or service "flows" from beginning of the value stream to the customer without waste.

Flow production: same as flow.

Gemba: a Japanese word meaning "real place," where action takes place— a shop floor or work areas.

Gemba walk: a walk carried out by a coach (a Lean sensei) and student or students to look for abnormal conditions, waste, or opportunities for improvement.

Heijunka: a method for leveling production for mix and volume.

Hoshin Kanri: a strategic decision-making tool used for policy deployment.

Internal setup: procedures that must be performed while the machine is stopped.

Ishikawa diagram: see fishbone diagram.

Jidoka: a device that stops production or equipment when a defective condition arises. Attention is drawn to this condition and the operator who stopped the production. The Jidoka system has faith in the operator who is trained for the job.

Just in time (JIT): originally developed by the Toyota Production System (TPS). JIT presupposes that all waste is eliminated from the production line, and only the inventory in the right quantity and at the right time is used for the production where the rate of production is exactly as required by the customer.

Kaikaku: a Japanese word meaning innovation or a radical breakthrough. Thus, Kaikaku requires radical thinking and takes more time in planning and implementation.

Kaizen: a Japanese word meaning change for the better or do good. It is a process of making continual improvements by everyone keeping in mind quality and safety.

Kaizen event: a short team-based improvement project; also called Kaizen blitz.

Kanban: means "sign board" or a label. It serves as an instruction for production and replenishment.

KCC: key critical characteristic.

KPC: key performance characteristic.

KPI: key performance indicator.

Lead time: a time required to move one piece from the time order is taken until it is shipped to the customer.

Line balancing: a technique where all operations are evenly balanced and staffing is also balanced to meet the takt time.

Malcolm Baldrige National Quality Award (MBNQA): an award given to the organization for achieving the highest quality standard.

Manufacturing resource planning (MRP II): an MRP but takes into consideration the capacity planning and finance requirement. It works out alternative production plans through the simulation tool.

Materials requirement planning (MRP): a computerized system of determining quantity and timing requirements for production and delivery of products for customers as well as suppliers. This is a PUSH production system.

Milk run: the routing of supply and delivery trucks/vehicles to make multiple pickups and deliveries at various locations to reduce transportation waste.

Muda: Japanese word for waste. It is an element that does not add value to the product or service. Also known as non-value-added activity carried out on a product or service that does not add value and the customer will not pay for it.

Mura: Japanese word for variability or unevenness.

Muri: a Japanese word for physical and mental strain or overburden.

One (single)-piece flow: practiced in the JIT system where one work piece flows from process to process to minimize waste.

Operational excellence (Opex): an element of organizational initiative that stresses the application of a variety of principles, systems, and tools toward the sustainable improvement of key performance metrics. This philosophy is based on continuous improvement, such as quality management system, Lean manufacturing, and Six Sigma. Operational excellence goes beyond the traditional methods of improvement and leads to a long-term change in organizational culture.

Overall equipment effectiveness (OEE): a product of the following key measures: (1) operational availability, (2) performance efficiency, and (3) first-pass yield quality.

PAIP: a process for performance analysis and improvement.

PCP: process control plan.

PDCA: plan, do, check, and act cycle for continual improvement.

PFMEA: process failure mode and effect analysis.

Point-of-use storage (POUS): storing or keeping materials, tools, information, and items near to where they are used.

Poka yoke: also known as mistake proofing. "Poka" in Japanese means inadvertent mistake, and "yoke" means prevention. These can be simple low-cost devices to sophisticated electro mechanical devices to prevent production of defective product.

Process: sequence of interdependent and linked procedures which, at every stage, consume one or more resources (employee time, energy, machines, money) to convert inputs (data, material, parts, etc.) into outputs. These outputs then serve as inputs for the next stage until a known goal or end result is reached.

Product realization: the term used to describe the work that the organization goes through to develop, manufacture, and deliver the finished product or service to the customer.

Productivity: measured as an output for a given input. Productivity increase is critical to improving living standards.

Pull: alternatively known as pull production where the upstream supplier does not produce until the downstream customer signals the need.

Push: alternatively known as push production where the upstream supplier produces as much as it can without regard to the fact whether the downstream customer needs it or not.

QOS: quality operating system originally implemented by Ford. The methodology was established to measure the effectiveness of the quality system and to drive continuous improvement.

Risk priority number (RPN): in FMEA, RPN = severity × occurrence × detection.

Rolled throughput yield (RTY): a probability that a single piece will pass through all production steps without a single defect.

Shadow board: a board where each tool has a place and in which tools are missing.

Single minute exchange of dies (SMED): a group of techniques developed by Shiego Shingo for the changeover of production equipment in less than 10 minutes.

SIPOC: a process identification where the requirements for supplier, input, process steps, output, and customer are defined.

Six Sigma: a set of tools and techniques for process improvement. It is originally developed by Motorola in 1981.

Spaghetti diagram: a diagram showing the layout and flow of information, material, and people in a work area. It is generally used to highlight motion and transportation waste.

Standard work: an accurate description of every process step specifying takt time, cycle time, minimum inventory needed, and sequence of each process step. The entire process is carried out with minimum human motion and other wastes.

Statistical process control (SPC): quality control where process variations are measures and controlled.

Supermarket: part storage before they go to the next operation. The parts are managed using minimum and maximum inventory levels.

Sustainability: continued development or growth, without significant deterioration of the environment and depletion of natural resources on which human well-being depends.

SWOT: stands for strength, weakness, opportunity, threat. Strength and weakness analysis guides us to identify the positives and negatives inside your organization (S-W), while opportunity and threat analysis guides us to identify positives and negatives outside of it. Developing a full SWOT analysis can help with strategic planning and decision making.

System: American Society for Quality (ASQ) defines system as "a group of interdependent processes and people that together perform a common mission."

Takt time: available production time divided by the rate of the customer demand.

Total productive maintenance (TPM): a system to ensure that every production process machine is able to perform its required tasks such that production is not interrupted.

Total quality management (TQM): a management approach that originated in the 1950s. The TQM culture requires quality in all aspects of the company's operations, with processes being done right the first time and defects and waste eradicated from operations. To be successful in implementing TQM, an organization must concentrate on the eight key elements: (1) ethics, (2) integrity, (3) trust, (4) training, (5) teamwork, (6) leadership, (7) recognition, and (8) communication.

VAA: value-added activity.

Value: a capability provided to a customer at the right time at an appropriate price, and is defined by the customer.

Value stream: sequence of actions required to design, produce, and provide a specific good or service and along which information, materials, and worth flows.

Visual factory: is a term to describe how data and information are conveyed to a Lean manufacturing environment. Here time and resources dedicated to conveying information are a form of waste. By using visual methods, information is easily accessible to those who need it. Visual information makes the current status of all processes immediately apparent.

Work in process (WIP): incomplete product or services that are awaiting further processing.

Bibliography

Automotive Industry Action Group, *Measurement System Analysis*, 3rd ed., AIAG, Southfield, MI, 2001.

Automotive Industry Action Group, *Potential Failure Mode and Effects Analysis: FMEA*, 3rd ed., AIAG, Southfield, MI, 2001.

Automotive Industry Action Group, *Statistical Process Control*, 3rd ed., AIAG, Southfield, MI, 2005.

American Society for Quality, *Certified Manager of Quality/Organizational Excellence.* Westcott, R. T. (ed.), ASQ Quality Press Hardcover, Milwaukee, WI, 2013.

American Society for Quality, *Certified Quality Auditor*, 3rd ed., Russell, J. P., ASQ Quality Press, Milwaukee, WI, 2011.

American Society for Quality, *Certified Quality Engineer.* Borror, C. M. (ed.), ASQ Quality Press Hardcover, Milwaukee, WI, 2009.

American Society for Quality, *Foundations in Quality Learning Series.* Patterson J. V. (ed.), ASQ Quality Press, Milwaukee, WI, 2009.

Brassard, M., *The Memory Jogger Plus+®.*

Crosby, P., Crosby's 14 steps to improvement, *Quality Progress*, December 2005.

Crosby, P., *Quality Without Tears*, McGraw-Hill, New York, 1984.

Deming, W. E., *Out of the Crisis*, MIT Press, Cambridge, MA, 1986.

Deming, W. E., *The New Economics for Industry, Government, Education*, MIT Press, Cambridge, MA, 1993.

Galsworth, G., Visual-Lean Institute, 1997.

Gryna, F. M., Chua, R. C. H., and DeFeo, J. A., *Juran's Quality Planning and Analysis for Enterprise Quality*, 5th ed., McGraw-Hill, 2006.

Juran, J. M., *Juran on Quality by Design*, The Free Press of the McMillan Inc., New York, 1992.

Pyzdek, T., *Quality Engineering Handbook*, 1st ed., ASQ Quality Press, Milwaukee, WI, 1991.

Pyzdek, T., *The Six Sigma Handbook: A Complete Guide for Green Belts, Black Belts, and Managers at All Levels*, 2nd ed., McGraw-Hill, New York, 2003.

Shingo, S. and Dillon, A., *A Study of the Toyota Production System: From an Industrial Engineering Viewpoint*, Productivity Press, New York, 1989.

The Toyota Production System, www.toyota Georgetown.com/tps.asp.

Wheeler, D. J., *Advanced Topics in Statistical Process Control*, SPC Press, Knoxville, TN, 1995.

Womack, J. P., Gembawalks Lean Enterprise Institute, 2011.

Index

About the Author

**Suresh Patel** is a former technical director and operations excellence executive. He holds a BE degree in electrical engineering from M.S. University of Baroda, India, a master's degree in production technology from South Bank University, London, and an MBA degree from the University of Texas at Brownsville, Texas. He is qualified as a Certified Reliability Engineer, Certified Quality Engineer, and Certified Management Systems Auditor certified by the American Society for Quality.

In his long career spanning more than four decades, he has developed a wide range of products/processes and has helped in establishing six manufacturing plants in India and five U.S. plants in Mexico. Starting with India, his career path has enabled him to work in industries in the UK, Denmark, Belgium, Canada, U.S., China, Mexico, and Chile. His career has been enriched through holding key positions with MNCs like Gestetner, Motorola, United Lighting Technologies, Eaton Corporation, and Fiat Global.

Patel's practical expertise and interests include establishment of Business Excellence Strategies starting from product quality strategies, quality improvement tools deployment, and execution, leading to improvements in product/process delivery performance and reduction in product escapes and product/process variation through Lean Six Sigma and overall business excellence employing Leadership and Results "Triades" as defined in MBNQA USA. Patel's other interests include supply chain management, manufacturing management, and building technological capabilities in manufacturing firms.